Distributed Cooperative Model Predictive Control of Networked Systems

网络系统的分布式协同模型预测控制

Yuanyuan Zou · Shaoyuan Li

邹媛媛　李少远　著

化学工业出版社

·北京·

内容简介

本书聚焦分布式网络化系统的高效协同控制问题，基于事件触发技术给出分布式预测控制器的设计和子系统协同方法，针对不同的控制需求对控制方法进行改进，实现系统控制性能、计算资源、通信资源的折中。本书提出了一种基于时间/事件混合触发的分布式模型预测控制（DMPC）方法，并对其可行性和稳定性进行了详细分析。本书所提的事件触发分布式模型预测控制方法具有独到的创新性和实用性，可应用于生产制造、智能电网、城市交通等领域，有助于实现大规模网络化系统的高效协同控制，为网络化系统智能控制提供理论和技术支撑。

本书为英文版，可供自动化及相关领域科研人员阅读，也可以作为相关专业研究生教材。

图书在版编目（CIP）数据

网络系统的分布式协同模型预测控制＝Distributed Cooperative Model Predictive Control of Networked Systems：英文/邹媛媛，李少远著．—北京：化学工业出版社，2024.8

ISBN 978-7-122-45702-8

Ⅰ.①网… Ⅱ.①邹…②李… Ⅲ.①计算机网络-网络系统-预测控制-英文 Ⅳ.①TP393.03

中国国家版本馆 CIP 数据核字（2024）第 102494 号

责任编辑：万忻欣　宋　辉　　　装帧设计：王晓宇

责任校对：赵懿桐

出版发行：化学工业出版社

（北京市东城区青年湖南街 13 号　邮政编码 100011）

印　　装：北京天宇星印刷厂

710mm×1000mm　1/16　印张 11½　字数 195 千字

2024 年 9 月北京第 1 版第 1 次印刷

购书咨询：010-64518888　　　售后服务：010-64518899

网　　址：http：//www.cip.com.cn

凡购买本书，如有缺损质量问题，本社销售中心负责调换。

定　　价：88.00 元

Preface

Networked Control Systems (NCSs) are spatially distributed systems in which the communication between sensors, actuators, and controllers is carried through a shared band-limited communication network. Examples of such systems include cyber-enabled manufacturing, smart grids, and water and sewage networks. Assisted by the highly efficient communication technology in networks, NCSs have the potential to achieve a consistent behavior among multiple distributed subsystems through the coordination and optimization of local controllers.

Distributed Model Predictive Control (DMPC) for such a complex system has been attracted much attention during the past twenty years, and most literatures on DMPC require these subsystems to communicate with each other in a synchronous manner. However, due to constrained network resources, the controller cannot exchange information as frequently and massively as they could in theory. The design of DMPC should consider not only the control requirements but also communication resources. This book is inspired by the development of DMPC of networked systems to save computation and communication resources.

The major new contribution is to show how to design efficient DMPCs that can be coordinated asynchronously with the increasing effectiveness of event-triggering mechanism, and how to improve the event-triggered DMPC for different requirements, namely improvement of control performance, extension to interconnected networked systems, etc. In Chap. 1, we recall the main concepts and some fundamental results of model predictive control and event-triggered/self-triggered control for NCSs. Some existing results on the event-triggered/self-triggered mechanisms, stability of the overall closed-loop system under DMPC strategies are provided. The DMPC approaches for of NCSs are presented in Chaps. 2-4. In Chap. 2, the DMPC of NCSs with event-triggered computation has been studied. The triggering conditions are obtained to decrease the solving frequency of optimization problems for each subsystem and a decentralized event-triggered dual-mode DMPC strategy is proposed to reduce information exchanges with neighboring subsystems. To further reduce the energy consumption in communication, the DMPC approaches of NCSs with event-triggered communication are introduced in Chap. 3. In Chap. 4, a dynamic event-triggering condition is presented, and we show that a larger inter-execution time can be obtained and the trade-off between resource usage and control performance is achieved. More complex scenarios are considered in Chaps. 5-7. A mixed time/event-triggered DMPC algorithm for the wired/wireless NCSs is introduced to improve system control performance in Chap. 5, and its feasibility and stability are also detailed. To avoid predefined triggering conditions continuously checked, the self-triggered DMPC is considered in Chap. 6, where the next triggering instant is predetermined based on the past information available at a triggering instant. At last, the event-triggered DMPC for large-scale NCSs with dynamic coupling is introduced in Chap. 7, two event-triggering conditions are established and comparability constraints on the predictive states at the triggering instants are imposed into the optimization problem, then feasibility and stability of the overall system are analyzed, respectively.

This book tries to make the readers understand the asynchronous distributed MPC, and to give a guidance to readers for designing distributed MPC and applying it to different areas. This book would be useful for the graduated students who are interest in the distributed MPC, and the persons who are engaged in researching control theory in academic institutes, university, and control engineering fields.

Yuanyuan Zou

Shaoyuan Li

Acknowledgements

This book is an accumulation and fruit of 7-year research in the area of distributed model predictive control for networked systems. We would like to express our gratitude to all those who have helped us. We thank Prof. James Lam at the University of Hong Kong, for his discussions and instructions in NCS and Prof. Daniel W C Ho at City University of Hong Kong for providing suggestions in multi-agent systems. In addition, we express thanks to Prof. Yugang Niu, Dr. Xiaoxiao Mi, Dr. Yaru Yang, doctoral candidate Tiange Yang, Mr. Xu Su, Ms. Ling Lu, and Ms. Fenglin Yuan for their support during the research period at Shanghai Jiao Tong University and East China University of Science and Technology. We also would like to thank the reviewers for reviewing this book and for those providing assistance to get this manuscript published. Finally, we would like to thank the financial support from the National Key R&D Program of China 2018YFB1701101, National Natural Science Foundations under Grants 62173224, 61833012, 61773162, and 61374107.

Yuanyuan Zou

Shaoyuan Li

Contents

Abbreviations

CMPC	Centralized model predictive control
DMPC	Distributed model predictive control
FHOCP	Finite horizon optimal control problem
ISS	Input-to-state stability
LMPC	Lyapunov-based model predictive control
LP-MPC	Linear program model predictive control
MPC	Model predictive control
NCSs	Networked control systems
QP-MPC	Quadratic program model predictive control
RMPC	Robust model predictive control
SMPC	Stochastic model predictive control
TT-DMPC	Time-triggered distributed model predictive control

Chapter 1
Introduction

1.1 Networked Control System

With the intended objectives of computational cost reduction and global performance improvement, more involved communication techniques have been integrated into the control architecture designs, for large-scale systems such as smart grids, urban transportation systems, industrial automated guided vehicle systems, water distribution systems, petrochemical processes and irrigation systems, etc. The above practical necessity outlines the developing theory and applications of networked control systems (NCSs), which possess the basic framework with multiple actuators, sensors and controllers communicating through the band-limited network [1,2]. With respect to the control structure, a NCS usually contains many subsystems with wide geographical distribution and relatively independent control-loops. In most cases, the controlled plant and the actuators among different subsystems are physically coupled. To achieve better control performance correspondingly, the specific controller of each subsystem conducts the manipulated variables for the actuators, utilizing the feedback measurements from its own sensors, as well as the controlling/sensing information from other subsystems. As a consequence, the subsystems possess complex coupling and correlation, in the concepts of both physical level and communicational level.

Compared to conventional centralized/decentralized control architectures, the NCSs take advantages of following aspects. On the one hand, the NCS

Y. Zou and S. Li, *Distributed Cooperative Model Predictive Control of Networked Systems*,
https://doi.org/10.1007/978-981-19-6084-0_1

structure provides a basic distributed realization for centralized control problems. Under the premise of achieving acceptable closed-loop results, this realization reduces the calculation burden and makes the system less sensitive to model mismatch and control errors [3]. Several studies within this framework have been carried out. For chemical processes with complex materials and energy interactions, the distributed hierarchical optimizations ensure the real-time capability of the real-time optimization and supervisory control [4]. In the secondary control of power systems, the distributed control admits the sparse communication network, while maintains the voltage and frequency of distributed generators in the acceptable ranges [5]. On the other hand, the NCS structure enables the cooperation of spatially distributed systems. With suitable communication protocol and sub-controllers, a NCS will outperform the conventional decentralized control one, in terms of robustness, efficiency and other operation performance, under flexible objectives. Along this routine, several distributed control strategies have been designed for plantwide chemical processes, islanded microgrid systems and multi-agent systems, etc. In the chemical engineering applications, the interconnections among different units were no longer considered as perturbation terms. The information from neighboring systems have intervened the local controller in terms of a feedforward manner [10]. In the islanded microgrid systems with intermittent renewable energy and diverse types of loads, the NCS techniques have been utilized in the energy-management, which maintains the system-wide supply demand balance and reduces operation cost [6-9]. Another special class of NCS applications that should be highlighted is the multi-agent system. Each subsystem has independent dynamics, while the coupling constraints exist among subsystems w. r. t. consensus, formation, obstacles avoidance and other global tasks. Many nontrivial control and cooperation strategies have been developed, toward the practically appealing scenarios including flexible manufacturing, forest fire monitoring, suspect tracking and military investigation etc. [11-15].

Broadly speaking, the control strategies utilized in the NCSs are in the distinct area of control theory according to the specific demands. For instance, the control Lyapunov function approach has been investigated to define and solve the cooperation goals e. g. formation of constrained NCSs. In this synthesis, a global Lyapunov function associated with the local Lyapunov functions of subsystems is constructed, and the coordination strategy can further be developed to achieve monotonic decreasing of the Lyapunov function [16]. Other methods based on Lyapunov techniques have also been developed [18,19], in which the Lyapunov/Riccati equations have been frequently combined with the state feedback control designs, considering various NCSs with different topologies, communication availability or individual dynamics, see e. g. [20-22,25]. The output feedback control methods have also been developed to achieve output tracking synchronization of NCSs with unmeasured states. The local algorithm collects the subsystem output as well as outputs from neighbors to estimate the local state distributively, and the local input is derived without constraints. In this area, the researchers have studied many scenarios including the Markov jump behaviors, internal communication delay or external disturbances involved in the dynamics, see e. g. [23,24]. Another mainstream design tool is the passivity concept that admits more systematical constructions of Lyapunov functions. The passive systems belong to a special kind of dissipative systems in which the amount of the stored energy is not more than the externally supplied energy. A broad class of feedback control laws can be designed within this concept, such that the global NCS inherits the passivity under specific interconnections; and as a consequence, the formation stability and tracking synchronization can be achieved, see e. g. [26-29]. Among the vast literature, model predictive control (MPC) scheme aligns well in the NCSs. This research avenue is the so-called distributed model predictive control (DMPC). The complex changeable global tasks are essentially formulated as the optimization cost, changeable physical constraints and/or cooperation constraints, which are the basic ingredients of the online constrained optimization problem of certain DMPC. The Lyapunov or passivity concepts are essentially integrated as the terminal constraints, terminal penalty terms or convergence constraints that qualitatively characterize the closed-loop behaviors.

In general, most of the existing control strategies in the NCSs researches have been relying on an ideal assumption, in which the information interactions are assumed to be conducted synchronously, under a global clock without delay and packet dropout. However, the real NCSs applications usually possess following characteristics that should be carefully considered.

- The sensors and other basic electronic units in the NCS usually contain different sampling period and dynamics. Therefore, it is difficult to ensure that the measurements would be collected synchronously.
- The time-scale separation exists among the dynamic characteristics of different subsystems, which is a common phenomenon in chemical processes. The synchronous cooperation strategy will enforce the consistent sampling period of the subsystems with slow/fast dynamics, and will further cause redundancy in both calculation and communication.
- Some synchronously computational and communicational efforts may be unnecessary, when the system state encounters slight disturbances or approaches the target steady-state. This feature deserves special attention, according to the coupling mechanism, uncertainty feature or steady-state phase duration of the NCS.
- The bandwidth, transfer rate and loading capability of the shared network are limited. Delay and packet dropouts may occur under these circumstances, for large-scale NCSs with immense amounts of information interactions.

To reduce the resource consumption for practical NCSs with aforementioned characteristics, recent researches have reported the multi-rate distributed control [30, 31], event-trigger distributed control [32-35] and related asynchronous cooperation strategies. In the online implementation of each subsystem, the controller is executed depending on the local dynamic and/or a predesigned event-triggering condition. The closed-loop systems exhibit acceptable control performance while reducing the computational and communicational costs. Such mechanism introduces new technical difficulties into the controller design, which includes the following two aspects.

- The feasibility of the controller is influenced by an asynchronous communication scheme. The controlling/sensing information from neighboring subsystems may be unavailable in the present execution of the local controller. The estimated state is used instead of the actual state. The estimation error should be addressed carefully such that the control system will not violate the process constraints and the coordination constraints.

- The stability property of the global NCS is difficult to guarantee when the dynamics exhibit time-scale separation among subsystems. On the one hand, for the constrained NCS with online optimization demand, the implicit control rules complicate the analysis of local closed-loop results. On the other hand, the local stability of subsystems doesn't necessarily imply global stability. The different converging rates introduce multi-stage uncertainty and high complexity.

In this book, we focus on the theoretical developments for cooperation of NCSs, including controller designs and closed-loop analysis. The main methodologies are within the DMPC framework, with event-triggered mechanism for computational and communicational resources reduction. In the rest of this section, we will introduce basic concepts and notable progress of MPC in Sect. 1.2, followed by the DMPC approaches introduced in Sect. 1.3. The event-trigger control with an emphasis on the event-trigger DMPC approaches will be reviewed in Sect. 1.4. Finally, the outlines of this book will be given in Sect. 1.5.

1.2 Model Predictive Control

MPC originated in the 1970s, from a standpoint of realizing online optimal control for constrained industrial process [36,44]. The regulators of MPC are always presented as linear/nonlinear constrained optimization problems. For a specific plant, the dynamical relationship of the concerned state and input are mathematically described by the predictive model. In the optimization problem, the model associates the predicted states with the optimization inputs over a finite prediction horizon, which together with the process restrictions constitute the constraint set. The optimization index, intuitively, is the accumulated deviation value between predicted state/input trajectories and the expected targets, with the latter ones evolved from the control goal (e.g. steady-state pairs for stabilization, or pre-optimized trajectories for tracking demand). With this setup, the optimization problem is usually referred to as the finite horizon optimal control problem (FHOCP). In the online implementation, the FHOCP is repeatedly solved in a receding horizon framework, i.e. at each sampling instant, the initial conditions are assigned according to current measurements, and the future control sequence with finite steps of input is optimized. After calculation, the input element associated with the current sampling interval is subsequently applied to the actuators of the plant. The MPC framework provides an implementable solution of online multi-variable control with changeable indexes and constraints, which adds robustness and flexibility to the optimal control mechanism. In the following decades, many efforts have been put into this research avenue, accelerating the developments in theoretical results and their applications. From the 1970s to the 1980s, the linear matrix control [36], the generalized predictive control [37, 38] and the state space model-based predictive control [39] were further evolved from this framework, with different forms of predictive model. The formulation of the MPC regulators was essentially in the heuristic sense in early approaches. From the 1980s to the 1990s, many researches have focused on the quantitative performance analysis of the MPC closed-loop systems; thereby the weighting matrices in the control indexes can be designed toward the expected dynamic evolution. In 1993, combined with the progress of optimal control theory, an MPC algorithm with stability guarantee has been proposed in [40]. Subsequently, the qualitative design synthesis based on the FHOCP terminal conditions has been established, toward the feasibility of the optimization and the convergence of the controlled system. Readers are referred to the literature [41-43], monographs [45, 46] and the references therein for details.

Driven by the practical operation demands, several theoretical branches of the receding horizon control schemes have been formed. Robust model predictive control (RMPC) is specifically designed for systems with bounded uncertainty (e. g. model mismatch or disturbances). By involving the min-max optimization strategy or the control invariant set techniques, some notable RMPC methods restrict the optimized state/input trajectories inside tightened constraint sets. On this basis, the closed-loop trajectories under uncertainty are proved to be inside the original constraint sets, which further ensure the feasibility and robust stability, e. g. [47-50]. To avoid conservatism, the stochastic model predictive control (SMPC) theory has been established, which provides the probability description of the constraints and the realization of the regulators. These methods prevent the designed robust conditions from conservatism by reducing the influence of the extreme cases. Correspondingly, the resulted closed-loops meet the control requirements most of the time, e. g. [51-53]. Another systematic scheme for systems with uncertainty is the Lyapunov-based model predictive control (LMPC), see the monograph [66] and the references therein. In this scheme, the Lyapunov constraints are incorporated into the online optimization problem. The offline designed auxiliary Lyapunov controller has been utilized in these constraints to regulate the convergent behavior of the state evolution. The invariant set indicating closed-loop stability can be further concluded by analyzing the stable and robust property of the auxiliary Lyapunov controller. For persistent disturbances and/or model mismatch, the so-called offset-free MPC schemes receive continual attentions, and have been widely applied to industrial systems. Compared to the RMPC, these schemes put special emphasis on the tracking performance associated with the static error (i. e. the offset), and relaxing the constraint set when necessary. According to the forms of the optimization problems, offset-free MPC schemes are also referred to as Linear program-MPC (LP-MPC) or Quadratic program-MPC (QP-MPC). Many theoretical results have been concluded including model structure and parameter designs [54-57], multi-priority constraint handling [58], stability analysis [59-61], etc.

The changeable control tasks occur occasionally in common practice, which have algorithmic representations in variable optimization indexes or switchable control modes. The former situation may result from, for example, changes in product profit and energy consumption in industrial applications. The optimal steady-states will vary within operating ranges during online implementation. Therefore, many MPC formulations with generic terminal conditions have been developed. In [62, 63], the terminal penalty terms were added to the control indexes to replace the role of terminal constraints. In this scheme, the domain of attraction is enlarged, through which the FHOCP will be feasible for changeable setpoints. Other researches proposed the so-called tracking MPC [64]. The terminal steady-state is formulated as an equilibrium to be optimized. The terminal condition toward the setpoint is essentially tightened gradually, with the terminal equilibrium gradually approaching the (possibly changed) setpoint, upon receding horizon optimization [65]. For systems with complex physical/chemical mechanisms, sets of multi-linear models are usually characterized to approximate the global dynamics. The MPC methods with multiple predictive models have emerged in response to these systems with changeable operating point. Some researches have organized the models and their valid operating ranges through fuzzy rules. In the online optimization, the predictive model is determined by weighted summation according to the fuzzy rules (defined on current state), e. g. [59,67]. Some researches have reported the multiple local MPC schemes and the transition strategies, e. g. [68, 69]. Other researchers considered to involve all predictive models in the FHOCP. At each sampling instant, a feasible control sequence is optimized, with the predicted trajectories (based on every model) satisfying the constraints. Although this scheme introduces additional conservatism, the stability conclusion can be drawn by involving the global stabilization constraint in the FHOCP [70,71].

1.3 Distributed Model Predictive Control

Section 1.2 reviews the prominent developments of MPC, which in general are established in a centralized framework. In order to achieve online constrained optimization of large-scale NCSs, many efforts have been put on the new scheme named as the DMPC. In a DMPC framework, the FHOCPs are solved respectively by the subsystems' computational units. The tracking goals are formulated as the optimization functions. At each sampling instant, the measured/estimated values are assigned to the FHOCP initial condition constraint; the neighboring optimized values (from the last computation) are assigned to the model constraint and/or consensus constraint. The local control sequence over a finite prediction horizon is computed by solving the FHOCP. The first element of the sequence is applied to the subsystem actuators, while the whole sequence is transmitted to the neighboring subsystems through the network if necessary.

In the conventional DMPC schemes for NCSs, all subsystems are sampled in accord with a global clock; the DMPC optimization problems are solved and the controlling/sensing information is communicated in a synchronous periodic manner. Focus on the specific concerns generated from certain applications, a lot of notable DMPC methods have been proposed with emphasis on different types of the network structures (e. g. neighboring or global communications), different modes of the information interactions (e. g. iterative, non-iterative, serial or parallel interactions), and different objectives of the FHOCPs (e. g. cooperative or non-cooperative optimizations). In the past decades, the main topics of the DMPC systems include the constraint satisfaction, global stability, optimality analysis and performance improvement, under coupling dynamics, combined constraints, disturbances or model mismatch. In the rest sections, we will review the classic DMPC methods, classified by the applicable NCSs with or without coupling dynamics among subsystems.

(1) DMPC for NCSs with decoupled dynamics

The NCS with decoupled dynamics refers to a group of spatially distributed subsystems without physical correlations among sub-plants, and with information interactions among sub-controllers. The control goals of these systems can be interpreted as follow: completing specific tasks through optimization and coordination, under given constraints and shared resources. It is worth mentioning that, consensus is a common task for this type of NCSs. A NCS exhibiting consensus property indicates that the certain physical variables (e. g. position, speed, temperature, etc.) of all subsystems converge to a constant upon the closed-loop evolution. In order to achieve consensus behavior, the optimization index of a local DMPC usually contains a coordinating term related to the states of neighboring subsystems. In this way, the optimized inputs are associated with the neighboring optimization results indirectly, which further enables the coordination of the subsystems through information interactions. In [72], a cooperative DMPC strategy has been proposed, for nonlinear NCSs that are dynamically decoupled. The optimization index of the sub-controller includes a tracking term designed for local dynamics, and a deviation term w. r. t the local state and the neighboring state. A consensus constraint restricting the bias between predicted states and observed states is added in the FHOCP, in order to achieve stability w. r. t. a tracked virtual leadership. The DMPC method proposed in [73] utilizes a serial interaction mode to serve the sequential optimizations among subsystems. The consensus is achieved by adding a coordination term respectively. Considering dynamically decoupled NCSs with bounded disturbances, a robust dual-mode DMPC has been proposed in [74]. A robust constraint is involved in the FHOCP, with the upper-bound varying in a contractive sense to achieve consensus with uncertainty. In [75], the above work was extended for NCSs with communicational uncertainty e. g. delay and packet dropouts. A waiting mechanism has been introduced into the DMPC. The sampling instant and coordination matrices are suitably defined in the FHOCP according to the delay characteristic, to achieve robust stability w. r. t. the consistent point. Considering nonlinear NCSs with fixed communicational delay, a coordinating index has been studied in [76], which depends on the delayed information from neighboring subsystems. The convergence results have been conducted based on the input-to-state stability (ISS) and the small gain condition.

The multi-agent systems belong to a special class of NCSs with decoupled dynamics. The formation problem of the multi-agent system can be regarded as the generalization of the consensus problem. A multi-agent system possesses a stable formation refers to that all subsystems maintain the consistent speed and the fixed relative positions. In practice, a challenging problem to be explored is how to ensure stable formation and avoid collisions among subsystems. In [77], the coordinating index, consensus constraint and avoidance constraint associated with the local state and neighboring state have been systematically given, for linear discrete-time multi-agent systems. However, the additional computation burden is inevitable due to the non-convex algorithm constraints. To accommodate to real-time applications, a robust DMPC strategy has been proposed for nonlinear online optimization, with the model parameters integrated as constraints [78]. In [79], by involving the feedback linearization technique, the FHOCP is transferred a mixed-integer quadratic programming problem. A new branch-and-bound-based algorithm has been proposed to loosen the avoidance constraint in [77], such that the computational burden can be reduced. Other important consensus problems explored in the DMPC literature include flocking or rendezvous through coordination (e. g. [80-82]).

(2) DMPC for NCSs with coupled dynamics

The NCSs with coupled dynamics exist in the smart grids, oil refining and petrochemical processes, water distribution systems, irrigation systems, etc. The characteristic of coupled dynamics refers to the interactions of mass/energy among subsystems. The evolution of local system is affected by the neighboring systems directly, which increases the control difficulty especially for strong coupling physical phenomenon. The DMPC design in this situation should put special emphasis on the treatment of coupling. Although the neighboring information (in last execution) has been transmitted to the local system, the prediction error of neighboring systems will affect the algorithm feasibility, and further degrades the global performance. A mainstream idea is to formulate the convergence constraint restricting deviations in FHOCP, such that the value of Lyapunov function decreases monotonically w. r. t. the closed-loop trajectory. In [83], a DMPC method has been proposed for nonlinear NCSs with weak coupling. The robust MPC techniques in [84, 85] have been introduced into the distributed control framework. The FHOCP contains an addictive consensus constraint with fix upper-bound, to restrict the bias between predicted states and estimated states. The closed-loop stability can be guaranteed, with a relative conservative condition associated the consensus constraint. Similar strategy has been extended from [83], for NCSs with dynamic coupling as well as input coupling [86]. In the following years, many efforts have been put on the new DMPC methods based on [83], with less conservatism. The DMPC in [87] contains a monotonically decreasing upper-bound of the consensus constraint, which enlarges the attraction domain to some extent. On this ground, the sufficient condition for feasibility and stability has been given. In [88], the contraction theory in [89] has been utilized to estimate the prediction error. A contractive constraint with a time-varying upper-bound is further constructed in the FHOCP, as the consensus constraint. This DMPC possesses an enlarged initial feasible set, and a less conservative condition of the applicable coupling characteristic of the NCS. From the aspect of information interaction, the aforementioned methods are formulated in a non-iterative fashion. For the current execution of local DMPC, part of the prediction error is attributed to the untimely information from the neighboring subsystems. Therefore, some researches (e. g. [90]) have reported the iterative DMPC approaches. At each sampling instant, the interactive influence of prediction errors is weakened by executing the FHOCPs and communicating the optimized results iteratively, until the errors of all subsystems are less than the threshold. In [91], an iterative DMPC accounting for global coordination performance has been proposed. Based on the convex optimization theory, an iterative equation for the communicated inputs is given. With this setup, the closed-loop performance of the DMPC systems achieve the global optimality. In [92], the tube-based robust MPC method is combined with the iterative DMPC. The interactive prediction errors are treated as uncertainties, which are bounded through the iterative execution. Based on [92], the parameterized terminal set has been developed, for interactions with large amplitude owing to the large state/input constraint sets [93].

Improving global performance with available communicational resources has been highlighted for decades in the DMPC researches. In general, expanding the coordination degree of subsystems with more extensive information interactions, will improve the performance of DMPC systems. Meanwhile, the consumption and the structural complexity of the network are the accompanied issues to be addressed. In the non-cooperative DMPC schemes [83, 87, 88], the control index of each FHOCP is only interrelated with the local tracking goal. The local DMPC receives the neighboring information that directly affects its dynamic prediction. In the cooperative DMPC schemes based on neighboring optimization [94-96], the control index considers the tracking goals of local subsystem as well as neighboring subsystems. The coordination degree has increased and the closed-loop performance has been improved. However, the FHOCP should take into account the information of subsystems that affects its neighborhood directly. In the cooperative DMPC schemes based on global information [90, 91], the control index of each FHOCP accounts for global tracking goal. These methods have the potential to achieve optimal performance theoretically, but require global information at each execution, which are deeply relying on the network resources.

1.4 Event Triggered Control

As is anticipated in the previews sections, the NCSs in real applications usually exhibit time-scale separation of the dynamics among subsystems. Moreover, in some of the cases, the computational and communicational resources are limited under the given control structure and the shared network. The trade-off between global performance and computational/communicational consumption has been an important research interest. This motivated the new scheme named as event-triggered control [32-35, 112, 113]. The basic concept of this scheme is to determine the instant of controller execution or information communication, according to the monitored state or the associated performance index. In this way, the computational burden and the communicational burden (for NCS cases) can be released, while closed-loop performance is acceptable w. r. t. the normal time-triggered situation.

Generally speaking, the core element of the event-triggered control design is the construction of the triggering condition. A common strategy is to analyze the system performance (such as Lyapunov stability [97-99], ISS [100-102], dissipativity [103, 104], L_2-gain stability [105-107]), with the latest states and the last computed input sequences. The event-triggering conditions are further established based on the dynamic behavior of the system according to the analysis results. In the online implementation, the controller updates the input sequence to improve performance once the triggering condition is satisfied; otherwise, the respective input element of the last computed sequence is applied. Some strategies considered to define the triggering condition directly. Subsequently, the sufficient conditions ensuring closed loop performance have been given resorting to the robust techniques [108, 109]. The other special type of condition is the self-triggered condition (e. g. [110, 111]). The next execution instant is conducted in an open-loop manner, which depends only on the current information including state and optimized input. Within this synthesis. it's unnecessary to monitor the state and to detect the triggering condition at every sampling instant. In the past decades, to align in the multiple applications, the event triggered scheme has been integrated into different types of control strategies, e. g. event-based PID control [112], event-triggered state-feedback control [114-116], event-triggered output-feedback control [117-119], event-triggered model predictive control and self-triggered model predictive control. We will review the notable results of the latter two methods in the rest of this section.

(1) Event-triggered MPC

Taking advantage of the receding horizon optimization mechanism, the eventtriggered MPC can utilize the open-loop optimized input sequence during the interval of two adjacent triggering instants. Focusing on the online optimization of NCSs with limited resources, many event-triggered MPC algorithms have been proposed. In [120], the event triggering condition depending on itself has been developed, for linear NCSs with decoupled dynamics. The FHOCP contains a consensus constraint defined on the predicted state and estimated state at the triggering instant. The close-loop analysis indicates that the method achieves feasibility and local input-to-state stability. In [121], this work has been improved by constructing a triggering condition defined on the local state and neighboring information. The global input-to-state stability can further be guaranteed. For nonlinear NCSs with coupled dynamics and in the presence of bounded disturbances, the self-triggered and event-triggered rules have been developed respectively, which are relying on the local state only. An eventtriggered DMPC algorithm has been proposed, with a stabilization constraint included in the FHOCP to ensure a decrease of the candidate Lyapunov function [122]. A similar problem has been addressed in [123], utilizing a periodic monitor mechanism. The event-triggering condition has been derived first, which does not involve the neighboring information. On this basis, the self-triggered condition has been constructed associated with the bounds of uncertainties, which significantly reduces the calculation and communication burden.

The event-triggered predictive control and optimization for NCSs with coupled dynamics is still a challenging topic to be explored. The main difficulty lies in ensuring feasibility and convergence, in the presence of dynamical interference and asynchronously obtained information. Some researches have reported the event-based communication conditions for iterative DMPC (e.g. [124,125]). The parallel optimization handling interactive influence is computationally realized by the iterative scheme. The influence of neighboring information on the local optimization results is quantified by sensitive analysis. The event-based communication conditions are further established to characterize the communicating actions among subsystems at each iterative. Other researches (e.g. [126]) have proposed the event-triggered robust MPC. The interactive correlations among subsystems are regarded as disturbances. These approaches are essentially accomplished in a decentralized framework, which exhibits conservatism w.r.t the applicable NCS network structures.

(2) Self-triggered MPC

In self-triggered MPC, the triggering condition can be derived from the analysis results of system performance, similarly. Different from the event-triggered formulation, the self-triggering condition is only defined on the estimated/measured state and the optimized input sequence obtained at the current triggering instant. Thus, the next triggering instant can be pre-defined [120,127,128]. In these methods, the self-triggering conditions are essentially independent of the MPC formulation, which is incompatible with the qualitative design synthesis and can not provide an priori performance guarantee. Recent advances in the selftriggered MPC have prominent progress on the combined design of the triggering mechanism and the FHOCP (e.g. [129,130]). The triggering instant is formulated as an optimization variable, i.e. by solving the FHOCP the optimal input sequence and the next triggering instant are obtained at once. In [129], the optimization objective is to minimize the number of executions while guaranteeing a priori chosen performance levels w.r.t. the convergence and constraint satisfaction. In order to achieve optimality, the formulation of the current FHOCP is relevant to the future optimization results (at the next triggering instant). This introduces additional difficulty into the computational realization. In [131], the future closed-loop optimization index has been approximated to an open-loop optimization index with a penalty term, thus the FHOCP can be realized. Moreover, to cope with uncertainty, the tube-based MPC techniques are utilized to achieve robust constraint satisfaction and asymptotic stability. A similar problem has been solved utilizing the Min-Max techniques in the self-triggered RMPC proposed by [132]. In [133], the self-triggered mechanism has been integrated into the stochastic MPC for linear systems with probabilistic constraints. This method improves the performance with relatively non-conservative algorithm conditions w. r. t. the self-triggered RMPC cases.

For NCSs controlled with MPC, the self-triggered mechanism brings flexibility to both control and communication. In [130], a centralized self-triggered MPC has been proposed for NCSs with the spatially distributed actuators and sensors. The optimization function in the FHOCP indicates the comprehensive cost of tracking performance and computational consumption. This method has been extended to the multi-loop self-triggered MPC in [134]. In the analysis part, the schedulability has been studied, which implies that the method achieves conflict free transmissions on the shared network. In [135], a novel optimization function has been proposed, which balances the performance and communication cost through an intuitive form. The adaptation of the intervals of triggers can be subsequently realized through optimization, w. r. t. the defined communication cost. Finally, the asymptotic stability of the resulting closed-loop system has been analyzed. It's worth mentioning that the closed-loop analysis of aforementioned methods is still oriented toward the centralized MPC (CMPC) systems. The research avenue about self-triggered distributed MPC is still in the initial stage.

1.5 Outline of This Book

This book focuses on the design of distributed cooperative MPC of NCSs to achieve the trade-off between global performance and computational/communicational consumption. The structure of the book is organized as follows. In this chapter, the properties of NCSs, the basic idea of MPC and DMPC are briefly introduced; an overview of the literature on DMPC for NCSs as well as the event-triggered/selftriggered MPC for NCSs are presented.

The DMPC approaches for NCSs are presented in Chaps. 2, 3, and 4. Chapt. 2 introduces two basic DMPC approaches of NCSs with event-triggered computation, where the FHOCP of each subsystem is solved at the triggering time instants and the information is exchanged periodically. A DMPC with cooperative event-triggered computation mechanism is first developed with a triggering condition considering the information received from neighboring subsystems, and a DMPC with decentralized event-triggered computation mechanism is then presented with a triggering condition only involving the information of the subsystem itself. We show that both triggering mechanisms ensure the ISS while saving the resource usage. In Chap. 3, a DMPC approach of NCSs with event-triggered communication is reported, where the subsystem solves an FHOCP and exchanges system information with its neighbors only if an event occurs. An event-triggering condition is built considering the decrease of the constructed Lyapunov function along with the triggering instants rather than the sampling instants, so as to decrease the triggering frequency and further reduce communication load. In Chap. 4, a dynamic variable considering the influence of neighboring subsystems is introduced in the design of the triggering condition. We show that a lower triggering frequency can be realized with the dynamic event triggering and the system performance is ensured. The algorithms, theoretical analysis, and examples are provided in these chapters.

More complex scenarios are considered in Chaps. 5, 6, and 7. The mixed wired wireless network has become a development trend for large-scale NCSs. Considering this communication environment, Chap. 5 investigates a mixed time/event-triggered dual-mode DMPC for NCSs, where an event-triggering condition of event-triggered subsystems is derived considering the influence of time/event-triggered communication patterns. For event-triggered control, the NCSs are generally continuously monitored and their preset triggering conditions are continuously checked, which is undesirable when the sampling cost is high. Considering this drawback, a selftriggered DMPC is considered in Chap. 6, where the next triggering instant is predetermined based on the past information available at a triggering instant, and sensor nodes can be in sleep during two successive triggering instants, thereby improving the effectiveness of monitoring and checking. Furthermore, considering that most practical large-scale processes usually consist of physically interconnected subsystems, Chap. 7 studies the design of the triggering mechanism to guarantee the performance of the event-triggered DMPC in presence of dynamic coupling. The algorithms, the-oretical analysis, and examples are also provided in these chapters.

References

1. Gupta, R. A., & Chow, M. Y. (2009). Networked control system: Overview and research trends. *IEEE Transactions on Industrial Electronics*, *57*(7), 2527-2535.
2. Hespanha, J. P., Naghshtabrizi, P., & Xu, Y. (2007). A survey of recent results in networked control systems. *Proceedings of the IEEE*, *95*(1), 138-162.
3. D'Andrea, R., & Dullerud, G. E. (2003). Distributed control design for spatially interconnected systems. *IEEE Transactions on Automatic Control*, *48*(9), 1478-1495.
4. Skogestad, S. (2004). Control structure design for complete chemical plants. *Computers and Chemical Engineering*, *28*(1-2), 219-234.
5. Xin, H., Qu, Z., Seuss, J., & Maknouninejad, A. (2010). A self-organizing strategy for power flow control of photovoltaic generators in a distribution network. *IEEE Transactions on Power Systems*, *26*(3), 1462-1473.
6. Zheng, Y., Li, S., & Tan, R. (2017). Distributed model predictive control for on-connected microgrid power management. *IEEE Transactions on Control Systems Technology*, *26*(3), 1028-1039.
7. Eddy, Y. F., Gooi, H. B., & Chen, S. X. (2014). Multi-agent system for distributed management of microgrids. *IEEE Transactions on Power Systems*, *30*(1), 24-34.
8. Du, Y., Wu, J., Li, S., Long, C., & Onori, S. (2019). Coordinated energy dispatch of autonomous microgrids with distributed MPC optimization. *IEEE Transactions on Industrial Informatics*, *15*(9), 5289-5298.
9. Du. Y., Wu. J., Li. S., Long, C., & Paschalidis, I. C. (2017). Distributed MPC for coordinated energy efficiency utilization in microgrid systems. *IEEE Transactions on Smart Grid*, *10*(2), 1781-1790.
10. Xu, S., & Bao, J. (2009). Distributed control of plantwide chemical processes. *Journal of Process Control*, *19*(10), 1671-1687.
11. Ren, W., & Beard, R. W. (2008). *Distributed consensus in multi-vehicle cooperative control*. Springer.
12. Wang, P., & Ding, B. (2014). Distributed RHC for tracking and formation of nonholonomic

multi-vehicle systems. *IEEE Transactions on Automatic Control*, *59*(6), 1439-1453.

13. Zhou, L., & Li, S. (2015). Distributed model predictive control for consensus of sampled-data multi-agent systems with double-integrator dynamics. *IET Control Theory and Applications*, *9*(12), 1774-1780.
14. Olfati-Saber, R., Fax, J. A., & Murray, R. M. (2007). Consensus and cooperation in networked multi-agent systems. *Proceedings of the IEEE*, *95*(1), 215-233.
15. Chen, S., Ho, D. W., Li, L., & Liu, M. (2014). Fault-tolerant consensus of multi-agent system with distributed adaptive protocol. *IEEE Transactions on Cybernetics*, *45*(10), 2142-2155.
16. Ögren, P., Egerstedt, M., & Hu, X. (2002). A control Lyapunov function approach to multi-agent coordination. *IEEE Transactions on Robotics and Automation*, *18*(5), 847-851.
17. Su, S., & Lin, Z. (2016). A multiple Lyapunov function approach to distributed synchronization control of multi-agent systems with switching directed communication topologies and unknown nonlinearities. *IEEE Transactions on Control of Network Systems*, *5*(1), 23-33.
18. Shi, G., & Hong, Y. (2009). Global target aggregation and state agreement of nonlinear multi-agent systems with switching topologies. *Automatica*, *45*(5), 1165-1175.
19. Hong, Y., Gao, L., Cheng, D., & Hu, J. (2007). Lyapunov-based approach to multiagent systems with switching jointly connected interconnection. *IEEE Transactions on Automatic Control*, *52*(5), 943-948.
20. Yue, D., Han, Q. L., & Peng, C. (2004). State feedback controller design of networked control systems. *IEEE Transactions on Circuits and Systems II: Express Briefs*, *51*(11), 640-644.
21. Ni, W., & Cheng, D. (2010). Leader-following consensus of multi-agent systems under fixed and switching topologies. *Systems and Control Letters*, *59*(3-4), 209-217.
22. Yu, W., Chen, G., Cao, M., & Kurths, J. (2009). Second-order consensus for multiagent systems with directed topologies and nonlinear dynamics. *IEEE Transactions on Systems, Man, and Cybernetics*, *40*(3), 881-891.
23. Wang, B. C., & Zhang, J. F. (2013). Distributed output feedback control of Markov jump multi-agent systems. *Automatica*, *49*(5), 1397-1402.
24. Liu, T., & Jiang, Z. P. (2013). Distributed output-feedback control of nonlinear multi-agent systems. *IEEE Transactions on Automatic Control*, *58*(11), 2912-2917.
25. Bidram, A., Davoudi, A., Lewis, F. L., & Guerrero, J. M. (2013). Distributed cooperative secondary control of microgrids using feedback linearization. *IEEE Transactions on Power Systems*, *28*(3), 3462-3470.
26. Bao, J., & Lee, P. L. (2007). *Process control: The passive systems approach*. Springer.
27. Arcak, M. (2007). Passivity as a design tool for group coordination. *IEEE Transactions on Automatic Control*, *52*(8), 1380-1390.
28. Zhu, Y., Qi, H., & Cheng, D. (2009). Synchronisation of a class of networked passive systems with switching topology. *International Journal of Control*, *82*(7), 1326-1333.

29. Rojas, O. J., Bao, J., & Lee, P. L. (2008). On dissipativity, passivity and dynamic operability of nonlinear processes. *Journal of Process Control*, *18*(5), 515-526.

30. Camponogara, E., & Talukdar, S. N. (2007). Distributed model predictive control: Synchronous and asynchronous computation. *IEEE Transactions on Systems, Man, and Cybernetics-Part A: Systems and Humans*, *37*(5), 732-745.

31. Zheng, C., Tippett, M. J., Bao, J., & Liu, J. (2013). Multirate dissipativity-based distributed MPC. In *2013 Australian control conference* (pp. 325-330). IEEE.

32. Hu, S., Yue, D., Yin, X., Xie, X., & Ma, Y. (2016). Adaptive event-triggered control for nonlinear discrete-time systems. *International Journal of Robust and Nonlinear Control*, *26*(18), 4104-4125.

33. Kia, S. S., Cortés, J., & Martinez, S. (2015). Distributed event-triggered communication for dynamic average consensus in networked systems. *Automatica*, *59*, 112-119.

34. Li, L., Ho, D. W., & Xu, S. (2014). A distributed event-triggered scheme for discrete-time multi-agent consensus with communication delays. *IET Control Theory and Applications*, *8*(10), 830-837.

35. Seyboth, G. S., Dimarogonas, D. V., & Johansson, K. H. (2013). Event-based broadcasting for multi-agent average consensus. *Automatica*, *49*(1), 245-252.

36. Cutler, C. R., & Ramaker, B. L. (1980). Dynamic matrix control-a computer control algorithm. In *Proceedings of Joint Automatic Control Conference* (Vol. 17, p. 72).

37. Clarke, D. W., Mohtadi, C., & Tuffs, P. S. (1987). Generalized predictive control-Part Ⅰ. The basic algorithm. *Automatica*, *23*(2), 137-148.

38. Clarke, D. W., Mohtadi, C., & Tuffs, P. S. (1987). Generalized predictive control-Part Ⅱ extensions and interpretations. *Automatica*, *23*(2), 149-160.

39. Richalet, J. (1993). Industrial applications of model based predictive control. *Automatica*, *29*(5), 1251-1274.

40. Rawlings. J. B., & Muske, K. R. (1993). The stability of constrained receding horizon control. *IEEE Transactions on Automatic Control*, *38*(10), 1512-1516.

41. Mayne, D. Q., Rawlings, J. B., Rao, C. V., & Scokaert, P. O. (2000). Constrained model predictive control: Stability and optimality. *Automatica*, *36*(6), 789-814.

42. Qin, S. J., & Badgwell, T. A. (2003). A survey of industrial model predictive control technology. *Control Engineering Practice*, *11*(7), 733-764.

43. Kwon, W. H., Han, S. H., & Ahn, C. K. (2004). Advances in nonlinear predictive control: A survey on stability and optimality. *International Journal of Control, Automation, and Systems*, *2*(1), 15-22.

44. Riehalet, J., Rault, A., Testud, J. L., & Papon, J. (1978). Model Predictive heuristic control: Applications to industrial Processes. *Automatica*, *14*(5), 413-428.

45. Camacho, E. F., & Bordons, C. (2007). *Model predictive control*. Springer.

46. Lars, G., & Jürgen, P. (2011). *Nonlinear model predictive control: Theory and algorithms*. Springer.

47. Marruedo, D. L., Alamo, T., & Camacho, E. F. (2002). Stability analysis of systems with bounded additive uncertainties based on invariant sets: Stability and feasibility of MPC. In *Proceedings of the 2002 American Control Conference* (Vol. 1, pp. 364-369). IEEE.

48. Mayne, D. Q., Seron, M. M., & Raković, S. V. (2005). Robust model predictive control of constrained linear systems with bounded disturbances. *Automatica*, *41* (2), 219-224. 219-224.

49. Limón, D., Alamo, T., Salas, F., & Camacho, E. F. (2006). Input to state stability of minmax MPC controllers for nonlinear systems with bounded uncertainties. *Automatica*, *42* (5), 797-803.

50. Li, D., Xi, Y., & Gao, F. (2013). Synthesis of dynamic output feedback RMPC with saturated inputs. *Automatica*, *49* (4), 949-954.

51. Cannon, M., Kouvaritakis, B., & Wu, X. (2009). Probabilistic constrained MPC for multiplicative and additive stochastic uncertainty. *IEEE Transactions on Automatic Control*, *54* (7), 1626-1632.

52. Cannon, M., Kouvaritakis, B., Raković, S. V., & Cheng, Q. (2010). Stochastic tubes in model predictive control with probabilistic constraints. *IEEE Transactions on Automatic Control*, *56* (1), 194-200.

53. Zou, Y., Lam, J., Niu, Y., & Li, D. (2015). Constrained predictive control synthesis for quantized systems with Markovian data loss. *Automatica*, *55*, 217-225.

54. Pannocchia, G., & Bemporad, A. (2007). Combined design of disturbance model and observer for offset-free model predictive control. *IEEE Transactions on Automatic Control*, *52* (6), 1048-1053.

55. Pannocchia, G., Gabiccini, M., & Artoni, A. (2015). Offset-free MPC explained: Novelties, subtleties, and applications. *IFAC-PapersOnLine*, *48* (23), 342-351.

56. Nikandrov, A., & Swartz, C. L. (2009). Sensitivity analysis of LP-MPC cascade control systems. *Journal of Process Control*, *19* (1), 16-24.

57. Marchetti, A. G., Ferramosca, A., & González, A. H. (2014). Steady-state target optimization designs for integrating real-time optimization and model predictive control. *Journal of Process Control*, *24* (1), 129-145.

58. Kerrigan, E. C., & Maciejowski, J. M. (2002). Designing model predictive controllers with prioritised constraints and objectives. In *Proceedings. IEEE International Symposium on Computer Aided Control System Design* (pp. 33-38). IEEE.

59. Zhang, T., Feng, G., & Zeng, X. J. (2009). Output tracking of constrained nonlinear processes with offset-free input-to-state stable fuzzy predictive control. *Automatica*, *45* (4), 900-909.

60. Zou T. (2012). Offset-free strategy by double-layered linear model predictive control. *Journal of Applied Mathematics*, 1927-1936.

61. Betti,G. ,Farina,M. ,& Scattolini,R. (2013). A robust MPC algorithm for offset-free tracking of constant reference signals. *IEEE Transactions on Automatic Control*, *58*(9),2394-2400.

62. Parisini, T. , & Zoppoli, R. (1995). A receding-horizon regulator for nonlinear systems and a neural approximation. *Automatica*, *31*(10),1443-1451.

63. Grüne,L. (2012). NMPC without terminal constraints. *IFAC Proceedings Volumes*, *45*(17), 1-13.

64. Limón,D. ,Alvarado,I. ,Alamo. T. ,& Camacho,E. F. (2008). MPC for tracking piecewise constant references for constrained linear systems. *Automatica*, *44*(9),2382-2387.

65. Ferramosca,A. ,Limon,D. ,Alvarado,I. ,Alamo,T. ,& Camacho. E. F. (2009). MPC for tracking with optimal closed-loop performance. *Automatica*, *8*(45),1975-1978.

66. Christofides,P. D. ,Liu,J. ,& Muñoz de la Peña,D. (2011). Lyapunov-based model predictive control. In *Networked and distributed predictive control* (pp. 13-45). Springer.

67. Roubos,J. A. ,Mollov,S. ,Babuška,R. ,& Verbruggen,H. B. (1999). Fuzzy model-based predictive control using Takagi-Sugeno models. *International Journal of Approximate Reasoning*, *22*(1-2),3-30.

68. Özkan, L. , Kothare, M. V. , & Georgakis, C. (2003). Control of a solution copolymerization reactor using multi-model predictive control. *Chemical Engineering Science*, *58*(7),1207-1221.

69. Özkan,L. ,& Kothare,M. V. (2006). Stability analysis of a multi-model predictive control algorithm with application to control of chemical reactors. *Journal of Process Control*, *16*(2), 81-90.

70. Badgwell,T. A. (1997). Robust model predictive control of stable linear systems. *International Journal of Control*, *68*(4),797-818.

71. Ferramosca. A. ,González. A. H. ,& Limon. D. (2017). Offset-free multi-model economic model predictive control for changing economic criterion. *Journal of Process Control*, *54*,1-13.

72. Gao,Y. ,Dai. L. ,Xia,Y. ,& Liu. Y. (2017). Distributed model predictive control for consensus of nonlinear second-order multi-agent systems. *International Journal of Robust and Nonlinear Control*, *27*(5),830-842.

73. Müller,M. A. ,Reble,M. ,& Allgöwer,F. (2012). Cooperative control of dynamically decoupled systems via distributed model predictive control. *International Journal of Robust and Nonlinear Control*, *22*(12),1376-1397.

74. Li,H. ,& Shi,Y. (2014). Robust distributed model predictive control of constrained continuous-time nonlinear systems: A robustness constraint approach. *IEEE Transactions on Automatic Control*, *59*(6),1673-1678.

75. Li,H. ,& Shi,Y. (2013). Robust distributed model predictive control of constrained continuous-time nonlinear systems: A robustness constraint approach. *IEEE Transactions on Automatic Control*, *59*(6),1673-1678.

76. Franco,E. ,Magni,L. ,Parisini,T. ,Polycarpou,M. M. ,& Raimondo,D. M. (2008). Cooperative

constrained control of distributed agents with nonlinear dynamics and delayed information exchange: A stabilizing receding-horizon approach. *IEEE Transactions on Automatic Control*, *53*(1), 324-338.

77. Wang, P., & Ding, B. (2014). A synthesis approach of distributed model predictive control for homogeneous multi-agent system with collision avoidance. *International Journal of Control*, *87*(1), 52-63.
78. Scholte, E., & Campbell, M. E. (2008). Robust nonlinear model predictive control with partial state information. *IEEE Transactions on Control Systems Technology*, *16*(4), 636-651.
79. Fukushima. H., Kon, K., & Matsuno, F. (2013). Model predictive formation control using branch-and-bound compatible with collision avoidance problems. *IEEE Transactions on Robotics*, *29*(5), 1308-1317.
80. Zhan, J., & Li, X. (2012). Flocking of multi-agent systems via model predictive control based on position-only measurements. *IEEE Transactions on Industrial Informatics*, *9*(1), 377-385.
81. Zhang, H. T., Cheng. Z., Chen, G., & Li, C. (2015). Model predictive flocking control for second-order multi-agent systems with input constraints. *IEEE Transactions on Circuits and Systems I: Regular Papers*, *62*(6), 1599-1606.
82. Zhou, L., & Li, S. (2017). Distributed model predictive control for multi-agent flocking via neighbor screening optimization. *International Journal of Robust and Nonlinear Control*, *27*(9), 1690-1705.
83. Dunbar, W. B. (2007). Distributed receding horizon control of dynamically coupled nonlinear systems. *IEEE Transactions on Automatic Control*, *52*(7), 1249-1263.
84. Marruedo, D. L., Alamo, T., & Camacho. E. F. (2002). Input-to-state stable MPC for constrained discrete-time nonlinear systems with bounded additive uncertainties. In *Proceedings of the 41 st IEEE Conference on Decision and Control, 2002* (Vol. 4, pp. 4619-4624). IEEE.
85. Chisci, L., Rossiter, J. A., & Zappa, G. (2001). Systems with persistent disturbances: Predictive control with restricted constraints. *Automatica*, *37*(7), 1019-1028.
86. Li, S., Zheng, Y., & Lin, Z. (2014). Impacted-region optimization for distributed model predictive control systems with constraints. *IEEE Transactions on Automation Science and Engineering*, *12*(4), 1447-1460.
87. Wang, P., & Ding, B. (2013). Distributed receding horizon control for dynamically coupled large scale systems. *IFAC Proceedings Volumes*, *46*(13), 254-259.
88. Long. Y., Liu, S., Xie, L., & Johansson, K. H. (2018). Distributed nonlinear model predictive control based on contraction theory. *International Journal of Robust and Nonlinear Control*, *28*(2), 492-503.
89. Forni, F., & Sepulchre, R. (2013). A differential Lyapunov framework for contraction analysis. *IEEE Transactions on Automatic Control*. *59*(3), 614-628.
90. Stewart, B. T., Venkat, A. N., Rawlings, J. B., Wright, S. J., & Pannocchia, G. (2010). Cooper-

ative distributed model predictive control. *Systems and Control Letters*, *59*(8), 460-469.

91. Venkat, A. N., Rawlings, J. B., & Wright, S. J. (2005). Stability and optimality of distributed model predictive control. In *Proceedings of the 44th IEEE Conference on Decision and Control* (pp. 6680-6685). IEEE.

92. Farina, M., & Scattolini, R. (2012). Distributed predictive control: A non-cooperative algorithm with neighbor-to-neighbor communication for linear systems. *Automatica*, *48*(6), 1088-1096.

93. Trodden, P. A., & Maestre, J. M. (2017). Distributed predictive control with minimization of mutual disturbances. *Automatica*, *77*, 31-43.

94. Zhang, Y., & Li, S. (2007). Networked model predictive control based on neighbourhood optimization for serially connected large-scale processes. *Journal of Process Control*, *17*(1), 37-50.

95. Zheng, Y., Li, S., & Wang, X. (2009). Distributed model predictive control for plant-wide hot-rolled strip laminar cooling process. *Journal of Process Control*, *19*(9), 1427-1437.

96. Li, S., Zhang, Y., & Zhu, Q. (2005). Nash-optimization enhanced distributed model predictive control applied to the Shell benchmark problem. *Information Sciences*, *170*(2-4), 329-349.

97. Branicky. M. S. (1998). Multiple Lyapunov functions and other analysis tools for switched and hybrid systems. *IEEE Transactions on Automatic Control*, *43*(4), 475-482.

98. Velasco, M., Martí, P., &Bini. E. (2009). On Lyapunov sampling for event-driven controllers. In *Proceedings of the 48h IEEE Conference on Decision and Control (CDC) held jointly with 2009 28th Chinese control conference* (pp. 6238-6243). IEEE.

99. Postoyan, R., Anta, A., Nešić, D., & Tabuada, P. (2011). A unifying Lyapunov-based framework for the event-triggered control of nonlinear systems. In *2011 50th IEEE conference on decision and control and European control conference* (pp. 2559-2564). IEEE.

100. Sontag, E. D. (2008). Input to state stability: Basic concepts and results. In *Nonlinear and optimal control theory* (pp. 163-220). Springer.

101. Tabuada, P. (2007). Event-triggered real-time scheduling of stabilizing control tasks. *IEEE Transactions on Automatic Control*, *52*(9), 1680-1685.

102. Mazo, M., & Tabuada, P. (2011). Decentralized event-triggered control over wireless sensor/actuator networks. *IEEE Transactions on Automatic Control*, *56*(10), 2456-2461.

103. Zhao, J., & Hill, D. J. (2008). Dissipativity theory for switched systems. *IEEE Transactions on Automatic Control*, *53*(4), 941-953.

104. Zhang, X. M., & Han, Q. L. (2015). A decentralized event-triggered dissipative control scheme for systems with multiple sensors to sample the system outputs. *IEEE Transactions on Cybernetics*, *46*(12), 2745-2757.

105. Van der Schaft, A. (2000). *L2-gain and passivity techniques in nonlinear control*. Springer.

106. Wang, X., & Lemmon, M. D. (2010). Event-triggering in distributed networked control systems. *IEEE Transactions on Automatic Control*, *56*(3), 586-601.

107. Donkers M, Heemels W. Donkers, M. C. F., & Heemels, W. P. M. H. (2010). Output-based

event-triggered control with guaranteed L_∞-gain and improved event-triggering. In *49th IEEE conference on decision and control* (*CDC*) (pp. 3246-3251). IEEE.

108. Lehmann, D., Henriksson, E., & Johansson, K. H. (2013). Event-triggered model predictive control of discrete-time linear systems subject to disturbances. In *2013 European Control Conference* (*ECC*) (pp. 1156-1161). IEEE.

109. Li, H., & Shi, Y. (2014). Event-triggered robust model predictive control of continuous-time nonlinear systems. *Automatica*, *50*(5), 1507-1513.

110. Anta, A., & Tabuada, P. (2008). Self-triggered stabilization of homogeneous control systems. In *2008 American control conference* (pp. 4129-4134). IEEE.

111. Heemels, W. P., Johansson, K. H., & Tabuada, P. (2012). An introduction to event-triggered and self-triggered control. In *2012 IEEE Conference on Decision and Control* (*CDC*) (pp. 3270-3285). IEEE.

112. Åarzén, K. E. (1999). A simple event-based PID controller. *IFAC Proceedings Volumes*, *32*(2), 8687-8692.

113. Åström, K. J., & Bernhardsson, B. (1999). Comparison of periodic and event based sampling for first-order stochastic systems. *IFAC Proceedings Volumes*, *32*(2), 5006-5011.

114. Lunze. J., & Lehmann, D. (2010). A state-feedback approach to event-based control. *Automatica*, *46*(1), 211-215.

115. Sahoo, A., Xu, H., & Jagannathan, S. (2015). Neural network-based event-triggered state feedback control of nonlinear continuous-time systems. *IEEE Transactions on Neural Networks and Learning Systems*, *27*(3), 497-509.

116. Yu, H., & Hao, F. (2016). Periodic event-triggered state-feedback control for discrete-time linear systems. *Journal of the Franklin Institute*, *353*(8), 1809-1828.

117. Zhang, J., & Feng, G. (2014). Event-driven observer-based output feedback control for linear systems. *Automatica*, *50*(7), 1852-1859.

118. Peng, C., & Zhang, J. (2015). Event-triggered output-feedback H_∞ control for networked control systems with time-varying sampling. *IET Control Theory and Applications*, *9*(9), 1384-1391.

119. Abdelrahim, M., Postoyan, R., Daafouz, J., & Nešić, D. (2015). Stabilization of nonlinear systems using event-triggered output feedback controllers. *IEEE Transactions on Automatic Control*, *61*(9), 2682-2687.

120. Hashimoto, K., Adachi, S., & Dimarogonas, D. V. (2015). Distributed aperiodic model predictive control for multi-agent systems. *IET Control Theory and Applications*, *9*(1), 10-20.

121. Zou, Y., Su, X., & Niu, Y. (2016). Event-triggered distributed predictive control for the cooperation of multi-agent systems. *IET Control Theory and Applications*, *11*(1), 10-16.

122. Varutti, P., Kern, B., Faulwasser, T., & Findeisen, R. (2009). Event-based model predictive control for networked control systems. In *Proceedings of the 48h IEEE Conference on Decision*

and Control (CDC) held jointly with 2009 28th Chinese Control Conference (pp. 567-572). IEEE.

123. Li, H. , Yan, W. , Shi, Y. , & Wang, Y. (2015). Periodic event-triggering in distributed receding horizon control of nonlinear systems. *Systems and Control Letters*, *86*, 16-23.

124. Groβ, D. , & Stursberg, O. (2015). A cooperative distributed MPC algorithm with event-based communication and parallel optimization. *IEEE Transactions on Control of Network Systems*, *3*(3), 275-285.

125. Groβ, D. (2015). *Distributed model predictive control with event-based communication*. University of Kassel.

126. Lu, L. , Zou, Y. , & Niu, Y. (2016). Event-triggered decentralized robust model predictive control for constrained large-scale interconnected systems. *Cogent Engineering*, *3*(1), 1127309.

127. Eqtami, A. , Heshmati-Alamdari, S. , Dimarogonas, D. V. , & Kyriakopoulos, K. J. (2013). Self-triggered model predictive control for nonholonomic systems. In *2013 European Control Conference (ECC)* (pp. 638-643). IEEE.

128. Hashimoto, K. , Adachi, S. , & Dimarogonas, D. V. (2016). Self-triggered model predictive control for nonlinear input-affine dynamical systems via adaptive control samples selection. *IEEE Transactions on Automatic Control*, *62*(1), 177-189.

129. Berglind, J. B. , Gommans, T. M. P. , & Heemels, W. P. M. H. (2012). Self-triggered MPC for constrained linear systems and quadratic costs. *IFAC Proceedings Volumes*, *45*(17), 342-348.

130. Henriksson, E. , Quevedo, D. E. , Sandberg, H. , & Johansson, K. H. (2012). Self-triggered model predictive control for network scheduling and control. *IFAC Proceedings Volumes*, *45*(15), 432-438.

131. Brunner, F. D. , Heemels, M. , & Allgöwer, F. (2016). Robust self-triggered MPC for constrained linear systems: A tube-based approach. *Automatica*, *72*, 73-83.

132. Liu, C. , Li, H. , Gao, J. , & Xu, D. (2018). Robust self-triggered min-max model predictive control for discrete-time nonlinear systems. *Automatica*, *89*, 333-339.

133. Dai, L. , Gao, Y. , Xie, L. , Johansson, K. H. , & Xia, Y. (2018). Stochastic self-triggered model predictive control for linear systems with probabilistic constraints. *Automatica*, *92*, 9-17.

134. Henriksson, E. , Quevedo. D. E. , Peters. E. G. , Sandberg, H. , & Johansson, K. H. (2015). Multiple-loop self-triggered model predictive control for network scheduling and control. *IEEE Transactions on Control Systems Technology*, *23*(6), 2167-2181.

135. Zhan, J. , Li. X. , & Jiang, Z. P. (2017. May). Self-triggered robust output feedback model predictive control of constrained linear systems. In *2017 American Control Conference (ACC)* (pp. 3066-3071). IEEE.

Chapter 2
DMPC of Networked Systems with Event-Triggered Computation

2.1 Introduction

Predictive control has attracted significant attention in the development of distributed control synthesis for large-scale NCSs, due to its ability to cope with the real-time changeable goals and constraints. The DMPC scheme is capable of eliminating the interactive effects among subsystems, by exchanging the predicted trajectories through the network. As is introduced in Chap. 1, the existing DMPC approaches for NCSs are mainly based on the time-triggered mechanism. The DMPC optimization problem is solved at each sampling time instant. This setting may be impractical for systems with constrained resources, and may also be unnecessary for slight uncertainty. In order to release the computation and communication burden for NCSs, the event-triggered DMPC scheme has been widely investigated in recent years. The methodology of these methods is to update the control action of each subsystem only when the event-triggering condition is satisfied and to remain constant until the next triggering time instant [1-5]. Although the event-triggered DMPC can be designed resorting to the analogous FHOCP formulation and stabilization tools of DMPC, the event-triggered mechanism directly results in the asynchronous feature of the information exchanges. This implies that the accumulated errors of the (predicted) neighboring states will influence the constraint satisfaction and the tracking performance of the NCS [6, 7]. Thus, how to design the event-triggering condition that ensures the algorithm feasibility and closed-loop stability, is an

Y. Zou and S. Li, *Distributed Cooperative Model Predictive Control of Networked Systems*,
https://doi.org/10.1007/978-981-19-6084-0_2

important topic to be discussed.

This chapter introduces two basic event-triggered DMPC approaches with the ensured closed-loop performance. The common framework of these two methods is to apply a trigger in the sensor-to-controller channel of each subsystem. Once the designed triggering condition is satisfied, the DMPC optimization problem is solved; the first element of the optimal input sequence is applied, and the optimization results are exchanged with neighbors. On the contrary, if the triggering condition is not satisfied, the pre-optimized input obtained at the last triggering instant is applied, and the current measured state or associated predicted state sequence are transmitted to the neighboring system if necessary. The main difference of the approaches lies in the formulation of the triggering conditions. In the first part, a cooperative eventtriggered computation mechanism is developed for NCSs with decoupled dynamics. The trigger needs to continuously monitor the information of local subsystem and neighboring subsystems, to realize the prompt optimization such that the candidate Lyapunov function is monotonically decreasing. In the latter part, this scheme is extended for NCSs with uncertainty, of which the disturbances and the weak coupling interaction (among subsystems) are considered through a perturbation term. The event-triggered computation mechanism is decentralized, since the triggering condition only depends on the information of subsystem itself. Once the states of two interactive subsystems both enter the pre-designed invariant set respectively, no information will be exchanged, further reducing the communication resources. The feasibility and stability are analyzed in both parts, and the simulations and comparison studies are conducted to verify the effectiveness of the approaches. The results of this chapter have been published in [8,9].

The remainder of this chapter is organized as follows. In Section 2. 2, the DMPC with a cooperative event-triggered computation mechanism is established. In Section 2. 3, the DMPC with a decentralized event-triggered computation mechanism is introduced. Section 2. 4 concludes the chapter.

Notation: Throughout this book, $\mathbb{R}^n$ denotes the real n dimensional Euclidean space; diag{$\cdots$} stands for a block-diagonal matrix; $\mathbb{N}$is the collection of all natural numbers; $\boldsymbol{I}_{n\times n}$ denotes the identity matrix with $n \times n$ dimension. The superscript T denotes the matrix transposition. Given a positive definite matrix $\boldsymbol{Q}$ and a column vector $\boldsymbol{x}=[x_1,\ldots,x_n]^{\mathrm{T}}$, $\overline{\lambda}(\boldsymbol{Q})$ and $\underline{\lambda}(\boldsymbol{Q})$ are the maximum eigenvalue and minimum eigenvalue of $\boldsymbol{Q}$, respectively; $\|\boldsymbol{x}\|_\infty=\max(|x_1|,\ldots,|x_n|)$ is the infinity norm of $\boldsymbol{x}$; $\|\boldsymbol{x}\|=\sqrt{\boldsymbol{x}^{\mathrm{T}}\boldsymbol{x}}$ and $\|\boldsymbol{x}\|_Q=\sqrt{\boldsymbol{x}^{\mathrm{T}}\boldsymbol{Q}\boldsymbol{x}}$ stand for the Euclidean norm and $\boldsymbol{Q}$-weighted norm of $\boldsymbol{x}$, respectively. A function $f(x)$ is said to be $\mathcal{K}$ function if it is continuous, strictly increasing and $f(0)=0$; a function $f(x)$ is said to be $\mathcal{K}_\infty$ function if it is a $\mathcal{K}$ function and $f(x)\to\infty$ as $x\to\infty$. A function $f(x)$ is said to be $\mathcal{K}$ function if it is continuous, strictly increasing and $f(0)=0$; a function $f(x)$ is said to be $\mathcal{K}_\infty$ function if it is a $\mathcal{K}$ function and $f(x)\to\infty$ as $x\to\infty$.

2.2 DMPC with a Cooperative Event-Triggered Computation Mechanism

2.2.1 *Optimization Problem Formulation*

Consider a linear discrete-time NCS, in which the dynamic of subsystem $\mathcal{A}_i$ $(i=1,\ldots,M)$ is given as

$$\boldsymbol{x}_i(k+1)=\boldsymbol{A}_i\boldsymbol{x}_i(k)+\boldsymbol{B}_i\boldsymbol{u}_i(k) \tag{2.1}$$

where $k\in N^+$ is the discrete time instant; $\boldsymbol{x}_i(k)\in\mathbb{R}^{n_i}$ and $\boldsymbol{u}_i(k)\in\mathbb{R}^{m_i}$ are the state and input vectors respectively; $\boldsymbol{A}_i\in\mathbb{R}^{n_i\times n_i}$ and $\boldsymbol{B}_i\in\mathbb{R}^{n_i\times m_i}$ are the constant matrices. The state and input are restricted by constraints $\boldsymbol{x}_i(k)\in\boldsymbol{X}_i$ and $\boldsymbol{u}_i(k)\in\boldsymbol{U}_i$, in which $\boldsymbol{X}_i$ and $\boldsymbol{U}_i$ are compact sets that contain the origin.

For specific subsystem $\mathcal{A}_i$ $(i=1,\ldots,M)$, the subsystem $\mathcal{A}_j$ $(j\neq i)$ is referred to as the neighbor of $\mathcal{A}_i$ if $\mathcal{A}_i$ and $\mathcal{A}_j$ have information interaction. The set of indexes of $\mathcal{A}_i$'s neighbors is denoted as $\mathcal{N}_i$. Considering the interaction is bi-directional, $j\in\mathcal{N}_i$ implies $i\in\mathcal{N}_j$.

The global state vector is defined as $\boldsymbol{x}(k)=[x_1^{\mathrm{T}}(k),\ldots,x_M^{\mathrm{T}}(k)]^{\mathrm{T}}$, and the input vector is defined as $\boldsymbol{u}(k)=[u_1^{\mathrm{T}}(k),\ldots u_M^{\mathrm{T}}(k)]^{\mathrm{T}}$, respectively. Combining with the subsystem model (2.1), the global NCS can be represented as

$$\boldsymbol{x}(k+1)=\boldsymbol{A}\boldsymbol{x}(k)+\boldsymbol{B}\boldsymbol{u}(k) \tag{2.2}$$

with $\boldsymbol{A}=\text{diag}\{A_1,\dots,A_M\}$ and $\boldsymbol{B}=\text{diag}\{B_1,\dots,B_M\}$.

The control goal of NCS (2.2) is to achieve global stability w. r. t. the set-point(i. e. the coordinate origin in this formulation)and to satisfy the state and input constraints of all subsystems. Fig. 2.1 depicts the structure of the proposed method. The DMPC controllers are designed distributively for each subsystem. A pre-designed event-based trigger is utilized in the sensor-to-controller channel to evaluate whether to execute the DMPC of each subsystem. At the sampling instant, the triggering condition is checked according to the local and neighboring information communicated through the network. The DMPC problem is solved only when this condition is satisfied, which leads to an aperiodic update of the optimized control sequence over the prediction horizon. In the rest of this section, we will show that by suitably formulating the triggering condition and the DMPC constraint set, the method ensures the desired performance with the reduced resources consumption.

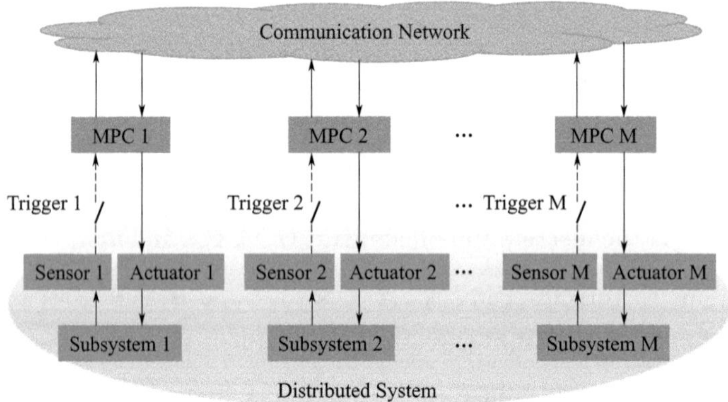

Fig. 2.1 Structure of event-triggered DMPC: the solid lines indicate that the information is transmitted periodically, and the dotted lines indicate that the information transmission is driven by event triggers

Let $k_i^d\ (d\in\mathbb{N})$ be the d -th triggering instant of the DMPC for subsystem $\mathcal{A}_i\ (i=1,\dots,M)$; and initialize k_i^d with $k_i^0=0$. At every triggering instant $k_i^d\ (d\in\mathbb{N})$, the DMPC solves following FHOCP to obtain the control sequence over prediction horizon $[k_i^d, k_i^d+N-1]$.

Problem 2.1 Let $\boldsymbol{u}_i(k_i^d+l\,|\,k_i^d)(l=0,\ldots,N-1)$ be the input sequence to be optimized; let $\boldsymbol{x}_i(k_i^d+l\,|\,k_i^d)(l=0,\ldots,N)$ be the predicted state sequence emanating from the measurement at instant k_i^d. The DMPC optimization problem to be solved at time k_i^d is given as

$$\min_{\boldsymbol{u}_i(k_i^d+l\,|\,k_i^d)} J_i(k_i^d) \tag{2.3}$$

subject to:

$$\boldsymbol{x}_i(k_i^d+l+1\,|\,k_i^d)=\boldsymbol{A}_i\boldsymbol{x}_i(k_i^d+l\,|\,k_i^d)+\boldsymbol{B}_i\boldsymbol{u}_i(k_i^d+l\,|\,k_i^d) \tag{2.4}$$

$$\hat{\boldsymbol{x}}_j(k_i^d+l+1\,|\,k_i^d)=\boldsymbol{A}_j\hat{\boldsymbol{x}}_j(k_i^d+l\,|\,k_i^d)+\boldsymbol{B}_j\hat{\boldsymbol{u}}_j(k_i^d+l\,|\,k_i^d) \tag{2.5}$$

$$\boldsymbol{x}_i(k_i^d+l\,|\,k_i^d)\in\boldsymbol{X}_i,\ l=1,\ldots,N-1 \tag{2.6}$$

$$\boldsymbol{u}_i(k_i^d+l\,|\,k_i^d)\in\boldsymbol{U}_i,\ l=0,\ldots,N-1 \tag{2.7}$$

$$\boldsymbol{x}_i(k_i^d+N\,|\,k_i^d)\in\boldsymbol{X}_{fi} \tag{2.8}$$

$$\|\boldsymbol{x}_i(k_i^d+l\,|\,k_i^d)-\hat{\boldsymbol{x}}_i(k_i^d+l\,|\,k_i^d)\|\leqslant\bar{\xi}_i(k_i^d),\ l=1,\ldots,N \tag{2.9}$$

with the control index defined as $J_i(k_i^d)=J_i^{\mathrm{E}}(k_i^d)+J_i^{\mathrm{F}}(k_i^d)$, and

$$J_i^{\mathrm{E}}(k_i^d)\triangleq\sum_{l=0}^{N-1}\left[\|\boldsymbol{x}_i(k_i^d+l\,|\,k_i^d)\|_{\boldsymbol{Q}_i}^2+\|\boldsymbol{u}_i(k_i^d+l\,|\,k_i^d)\|_{\boldsymbol{R}_i}^2\right]+\|\boldsymbol{x}_i(k_i^d+N\,|\,k_i^d)\|_{\boldsymbol{P}_i}^2$$

$$J_i^{\mathrm{F}}(k_i^d)\triangleq\sum_{l=0}^{N-1}\sum_{j\in\mathcal{N}_i}\|\boldsymbol{x}_i(k_i^d+l\,|\,k_i^d)-\hat{\boldsymbol{x}}_j(k_i^d+l\,|\,k_i^d)\|_{\boldsymbol{Q}_{ij}}^2$$

where $J_i^{\mathrm{E}}(k_i^d)$ is the local regulation term and $J_i^{\mathrm{F}}(k_i^d)$ is the cooperative term; N is the prediction horizon; $\boldsymbol{Q}_i$ and $\boldsymbol{R}_i$ are the positive definite weighting matrices; $\boldsymbol{P}_i$ is the terminal weighting matrix; $\boldsymbol{Q}_{ij}$ is the cooperation matrix; $\hat{\boldsymbol{x}}_j(k_i^d+l\,|\,k_i^d)$ and $\hat{\boldsymbol{u}}_j(k_i^d+l\,|\,k_i^d)$ are the estimated state and control input of neighbor $\mathcal{A}_j$; $\hat{\boldsymbol{x}}_i(k_i^d+l\,|\,k_i^d)$ is the estimated state generated by neighbor $\mathcal{A}_j$, based on the predicted state received from subsystem $\mathcal{A}_i$. The terminal predicted state $\boldsymbol{x}_i(k_i^d+N\,|\,k_i^d)$ is restricted by constraint (2.8), with the set $\boldsymbol{X}_{fi}$ defined as

$$\boldsymbol{X}_{fi}=\{\boldsymbol{x}_i\in\mathbb{R}^{n_i}:\|\boldsymbol{x}_i\|_{\boldsymbol{P}_i}^2\leqslant\varepsilon_i^2,\varepsilon_i>0\} \tag{2.10}$$

Constraint (2.9) restricts the difference between predicted state $\boldsymbol{x}_i(k_i^d+l\,|\,k_i^d)$ and estimated $\hat{\boldsymbol{x}}_j(k_i^d+l\,|\,k_i^d)$, where the upper bound $\bar{\xi}_i(k_i^d)$ is a scalar related to the triggering instant k_i^d.

Considering the asynchronous execution of subsystem $\mathcal{A}_i$ and its neighbor $\mathcal{A}_j$, the estimated state $\hat{\boldsymbol{x}}_j(k_i^d|k_i^d)$ and control input $\hat{\boldsymbol{u}}_j(k_i^d+l|k_i^d)$ have specific definitions according to the last triggering instant of $\mathcal{A}_j$, which are clarified as follows.

(1) For the case with triggering instant of $\mathcal{A}_j$ being k_i^d-1, there is

$$\begin{aligned}&\hat{\boldsymbol{x}}_j(k_i^d|k_i^d)=\boldsymbol{x}_j^*(k_i^d|k_i^d-1),\\&\hat{\boldsymbol{u}}_j(k_i^d+l|k_i^d)=\begin{cases}\boldsymbol{u}_j^*(k_i^d+l|k_i^d-1),l=0,\ldots,N-2\\\boldsymbol{K}_j\hat{\boldsymbol{x}}_j(k_i^d+l|k_i^d),l=N-1,\end{cases}\end{aligned}\tag{2.11}$$

where $\boldsymbol{x}_j^*(k_i^d|k_i^d-1)$ and $\boldsymbol{u}_j^*(k_i^d+l|k_i^d-1)$ denote the optimal predicted state and optimal control sequence of DMPC Problem 2.1 at instant k_i^d-1, w. r. t subsystem $\mathcal{A}_j$.

(2) For the case with triggering instant of $\mathcal{A}_j$ earlier than k_i^d-1, there is

$$\begin{aligned}&\hat{\boldsymbol{x}}_j(k_i^d|k_i^d)=\hat{\boldsymbol{x}}_j(k_i^d|k_i^d-1),\\&\hat{\boldsymbol{u}}_j(k_i^d+l|k_i^d)=\begin{cases}\overline{\boldsymbol{u}}_j(k_i^d+l|k_i^d-1),l=0,\ldots,N-2\\\boldsymbol{K}_j\hat{\boldsymbol{x}}_j(k_i^d+l|k_i^d),l=N-1\end{cases}\end{aligned}\tag{2.12}$$

where $\overline{\boldsymbol{x}}_j(k_i^d|k_i^d-1)$ and $\overline{\boldsymbol{u}}_j(k_i^d+l|k_i^d-1)$ denote the feasible predicted state and feasible control sequence of Problem 2.1 at instant k_i^d-1, w. r. t. subsystem $\mathcal{A}_j$. The construction of the feasible sequences is based on the optimized results at last triggering instants and the local state feedback control law $\boldsymbol{K}_i(i=1,\ldots M)$. Specifically, the parameters in the control index and the feedback control law $\boldsymbol{K}_i(i=1,\ldots,M)$ satisfy the following assumption.

Assumption 2.1 For subsystem $\mathcal{A}_i(i=1,\ldots,M)$, the feedback control law $\boldsymbol{K}_i$ satisfies $\boldsymbol{K}_i\boldsymbol{x}_i\in\boldsymbol{U}_i$. Moreover, there exists a constant $\varepsilon_i>0$, such that for all $\boldsymbol{x}_i\in\boldsymbol{X}_{fi}=\{\boldsymbol{x}_i\in\mathbb{R}^{n_i}:\|\boldsymbol{x}_i\|_{P_i}^2\leqslant\varepsilon_i^2\}$, there is

$$\begin{aligned}&(\boldsymbol{A}_i+\boldsymbol{B}_i\boldsymbol{K}_i)^{\mathrm{T}}\boldsymbol{P}_i(\boldsymbol{A}_i+\boldsymbol{B}_i\boldsymbol{K}_i)-\boldsymbol{P}_i+\overline{\boldsymbol{Q}}_i+\sum_{j\in\mathcal{N}_i}2(\boldsymbol{Q}_{ij}+\boldsymbol{Q}_{ji})\leqslant0\\&\overline{\boldsymbol{Q}}_i=\boldsymbol{Q}_i+\boldsymbol{K}_i^{\mathrm{T}}\boldsymbol{R}_i\boldsymbol{K}_i\end{aligned}\tag{2.13}$$

where $\boldsymbol{Q}_i$, $\boldsymbol{R}_i$, $\boldsymbol{P}_i$ are the positive definite symmetric matrices, and $\boldsymbol{Q}_{ij}$, $\boldsymbol{Q}_{ji}$ are the cooperation matrices utilized in Problem 2.1.

Remark 2.1 The cooperation term $\|\boldsymbol{x}_i(k_i^d+l|k_i^d)-\hat{\boldsymbol{x}}_j(k_i^d+l|k_i^d)\|_{\boldsymbol{Q}_{ij}}^2$ is formulated to regulate the consensus manner of subsystem $\mathcal{A}_i$ and its neighbors. In detail, the bias of the predicted states among subsystem $\mathcal{A}_i$ and its neighbors $\mathcal{A}_j(j\in\mathcal{N}_i)$ are penalized in the control index. The values of states are expected to achieve consensus for the NCSs (2.1), through receding horizon implementation. It's worth mentioning that the chosen of matrix $\boldsymbol{Q}_{ij}$ will significantly influence the cooperation behavior among subsystems. The readers are referred to references [6,7] for the design and discussion of $\boldsymbol{Q}_{ij}$ according to the specific model description and cooperation goals of the NCSs.

Remark 2.2 In this method, the determination of local feedback control law $\boldsymbol{K}_i$ for subsystem $\mathcal{A}_i$ relies not only on the weighting matrices $\boldsymbol{Q}_i$, $\boldsymbol{R}_i$, $\boldsymbol{P}_i$ as in [7], but also on the cooperation matrices $\boldsymbol{Q}_{ij}$ and $\boldsymbol{Q}_{ji}$. On this basis, the global ISS Lyapunov function can be derived and the convergence of the NCS can be guaranteed.

2.2.2 *Cooperative Event-Triggering Condition*

Section 2.2.1 gives the basic formulation of the DMPC problem. In this section, the event-triggering condition will be conducted, such that the closed-loop stability can be guaranteed while the communicational and computational costs can be reduced.

To proceed, we first construct the feasible control sequence between two consecutive triggering instants k_i^d and k_i^{d+1}, w.r.t. subsystem $\mathcal{A}_i$. Denote the optimal control sequence of Problem 2.1 at time instant k_i^d as $u_i^*(k_i^d+l|k_i^d)$; denote the related optimal state trajectory $\boldsymbol{x}_i^*(k_i^d+l|k_i^d)$ and the optimal cost as $J_i^*(k_i^d)$. For time interval $k\in[k_i^d,k_i^{d+1})$, the feasible control sequence $\bar{\boldsymbol{u}}_i(k+l|k)$ is defined as follows.

For $k=k_i^d+1$,

$$\bar{\boldsymbol{u}}_i(k+l|k)=\begin{cases}\boldsymbol{u}_i^*(k+l|k_i^d), l=0,\ldots,N-2\\ \boldsymbol{K}_i\bar{\boldsymbol{x}}_i(k+l|k), l=N-1\end{cases} \tag{2.14}$$

For $k_i^d+1<k<k_i^{d+1}$,

$$\bar{\boldsymbol{u}}_i(k+l|k)=\begin{cases}\bar{\boldsymbol{u}}_i(k+l|k-1), l=0,\ldots,N-2\\ \boldsymbol{K}_i\bar{\boldsymbol{x}}_i(k+l|k), l=N-1\end{cases} \tag{2.15}$$

where the state and input sequences $\bar{\boldsymbol{x}}_i(k+l|k)$ and $\bar{\boldsymbol{u}}_i(k+l|k)$ satisfy the subsystem model (2.2).

The global candidate Lyapunov function is the sum of subsystems' cost functions. To construct the triggering condition that guarantees the monotonically decreasing property of the function, we analyze the difference of $J(k+1)$ and $J(k)$, in which $J(k)=\sum_{i=1}^{M}\bar{J}_i(k)$ and $\bar{J}_i(k)$ denotes the cost value under control sequence (2.14) or (2.15).

$$\begin{aligned}
&\Delta J(k+1)\\
&=J(k+1)-J(k)\\
&=\sum_{i=1}^{M}\left\{\sum_{l=1}^{N-1}\sum_{j\in\mathcal{N}_i}[\|\bar{\boldsymbol{x}}_i(k+l|k+1)-\hat{\boldsymbol{x}}_j(k+l|k+1)\|_{\boldsymbol{Q}_{ij}}^2-\|\bar{\boldsymbol{x}}_i(k+l|k)-\right.\\
&\quad\hat{\boldsymbol{x}}_j(k+l|k)\|_{\boldsymbol{Q}_{ij}}^2]\\
&\quad-\|\boldsymbol{x}_i(k)\|_{\boldsymbol{Q}_i}^2-\|\boldsymbol{u}_i^*(k|k_i^d)\|_{\boldsymbol{R}_i}^2-\sum_{j\in\mathcal{N}_i}\|\boldsymbol{x}_i(k)-\hat{\boldsymbol{x}}_j(k|k)\|_{\boldsymbol{Q}_{ij}}^2\\
&\quad+\|\bar{\boldsymbol{x}}_i(k+N+1|k+1)\|_{\boldsymbol{P}_i}^2\\
&\quad-\|\bar{\boldsymbol{x}}_i(k+N|k)\|_{\boldsymbol{P}_i}^2+\|\bar{\boldsymbol{x}}_i(k+N|k+1)\|_{\boldsymbol{Q}_i}^2+\sum_{j\in\mathcal{N}_i}\|\bar{\boldsymbol{x}}_i(k+N|k+1)\\
&\quad\left.-\hat{\boldsymbol{x}}_j(k+N|k+1)\|_{\boldsymbol{Q}_{ij}}^2\right\}
\end{aligned} \tag{2.16}$$

Define $\Theta_{i1}\triangleq\|\bar{\boldsymbol{x}}_i(k+l|k+1)-\hat{\boldsymbol{x}}_j(k+l|k+1)\|_{\boldsymbol{Q}_{ij}}^2-\|\bar{\boldsymbol{x}}_i(k+l|k)-\hat{\boldsymbol{x}}_j(k+l|k)\|_{\boldsymbol{Q}_{ij}}^2$, $\Theta_{i2}\triangleq\|\bar{\boldsymbol{x}}_i(k+N+1|k+1)\|_{\boldsymbol{P}_i}^2-\|\bar{\boldsymbol{x}}_i(k+N|k)\|_{\boldsymbol{P}_i}^2+\|\bar{\boldsymbol{x}}_i(k+N|k+1)\|_{\boldsymbol{Q}_i}^2+\sum_{j\in\mathcal{N}_i}\|\bar{\boldsymbol{x}}_i(k+N|k+1)-\hat{\boldsymbol{x}}_j(k+N|k+1)\|_{\boldsymbol{Q}_{ij}}^2$. Noting $\bar{\boldsymbol{x}}_i(k+l|k+1)=\bar{\boldsymbol{x}}_i(k+l|k)$ and $\hat{\boldsymbol{x}}_j(k+l|k+1)=\bar{\boldsymbol{x}}_j(k+l|k)$, and as a result of the triangle inequality, there is

$$\begin{aligned}
\Theta_{i1}&=\|\bar{\boldsymbol{x}}_i(k+l|k)-\bar{\boldsymbol{x}}_j(k+l|k)\|_{\boldsymbol{Q}_{ij}}^2-\|\bar{\boldsymbol{x}}_i(k+l|k)-\hat{\boldsymbol{x}}_j(k+l|k)\|_{\boldsymbol{Q}_{ij}}^2\\
&\leqslant\bar{\lambda}(\boldsymbol{Q}_{ij})[2\|\bar{\boldsymbol{x}}_i(k+l|k)-\hat{\boldsymbol{x}}_j(k+l|k)\|\|\bar{\boldsymbol{x}}_j(k+l|k)-\hat{\boldsymbol{x}}_j(k+l|k)\|\\
&\quad+\|\bar{\boldsymbol{x}}_j(k+l|k)-\hat{\boldsymbol{x}}_j(k+l|k)\|^2]\\
&\leqslant\bar{\lambda}(\boldsymbol{Q}_{ij})[2(\xi_i(k+l|k)+\eta_i(k+l|k))\xi_j(k+l|k)+\xi_j^2(k+l|k)]
\end{aligned} \tag{2.17}$$

where $\xi_i(k+l\mid k)=\|\overline{x}_i(k+l|k)-\hat{x}_i(k+l|k)\|$, $\xi_j(k+l\mid k)=\|\overline{x}_j(k+l|k)-\hat{x}_j(k+l|k)\|$ and $\eta_i(k+l\mid k)=\|\hat{x}_i(k+l|k)-\hat{x}_j(k+l|k)\|$.

According to the symmetric relation among subsystems, there is

$$\begin{aligned}
&\sum_{i=1}^{M}\sum_{j\in\mathcal{N}_i}\|\overline{x}_i(k+N|k+1)-\hat{x}_j(k+N|k+1)\|_{Q_{ij}}^2\\
&\leqslant\sum_{i=1}^{M}\sum_{j\in\mathcal{N}_i}2\|\overline{x}_i(k+N|k)\|_{Q_{ij}}^2+\sum_{i=1}^{M}\sum_{j\in\mathcal{N}_i}2\|\overline{x}_j(k+N|k)\|_{Q_{ij}}^2\\
&=\sum_{i=1}^{M}\sum_{j\in\mathcal{N}_i}2\|\overline{x}_i(k+N|k)\|_{Q_{ij}+Q_{ji}}^2
\end{aligned}\tag{2.18}$$

Combining with (2.13), we further have

$$\begin{aligned}
\sum_{i=1}^{M}\Theta_{i2}&\leqslant\sum_{i=1}^{M}\{\|(\boldsymbol{A}_i+\boldsymbol{B}_i\boldsymbol{K}_i)\overline{x}_i(k+N|k)\|_{\boldsymbol{P}_i}^2-\|\overline{x}_i(k+N|k)\|_{\boldsymbol{P}_i}^2+\\
&\quad\|\overline{x}_i(k+N|k)\|_{\overline{Q}_i}^2\\
&\quad+\sum_{j\in\mathcal{N}_i}2\|\overline{x}_i(k+N|k)\|_{Q_{ij}+Q_{ji}}^2\Big\}\\
&=\sum_{i=1}^{M}\|\overline{x}_i(k+N|k)\|^2_{(\boldsymbol{A}_i+\boldsymbol{B}_i\boldsymbol{K}_i)^{\mathrm{T}}\boldsymbol{P}_i(\boldsymbol{A}_i+\boldsymbol{B}_i\boldsymbol{K}_i)-\boldsymbol{P}_i+\overline{Q}_i+\sum_{j\in\mathcal{N}_i}2(Q_{ij}+Q_{ji})}\\
&\leqslant 0
\end{aligned}\tag{2.19}$$

Substituting (2.17) and (2.19) into (2.16), it follows that

$$\Delta J(k+1)\leqslant\sum_{i=1}^{M}\{-\Xi_i+\Pi_i\}\tag{2.20}$$

with

$$\Xi_i=\|\boldsymbol{x}_i(k)\|_{Q_i}^2+\|\boldsymbol{u}_i^*(k|k_i^d)\|_{\boldsymbol{R}_i}^2+\sum_{j\in\mathcal{N}_i}\|\boldsymbol{x}_i(k)-\hat{x}_j(k|k)\|_{Q_{ij}}^2\tag{2.21}$$

$$\Pi_i=\sum_{l=1}^{N-1}\sum_{j\in\mathcal{N}_i}\overline{\lambda}(Q_{ij})[2(\xi_i(k+l|k)+\eta_i(k+l|k))\cdot\xi_j(k+l|k)+\xi_j^2(k+l|k)]\tag{2.22}$$

Let $\Pi_i\leqslant\sigma_i\,\Xi_i$ with triggering parameter $0<\sigma_i<1$, there is $-\Xi_i+\Pi_i\leqslant(\sigma_i-1)\Xi_i<0$. Finally, the event-triggering condition at time $k+1$ is derived as

$$\Pi_i>\sigma_i\,\Xi_i\tag{2.23}$$

The trade-off between the control performance and communicational/computational resources can be achieved, by tuning the value of triggering parameter σ_i in condition (2.24). It can be easily observed that, smaller σ_i leads to more frequent execution of Problem 2.1, which gets the better control performance. Particularly, when $\sigma_i=0$, the proposed DMPC can be regarded as a time-triggered DMPC (TT-DMPC). On the contrary, larger σ_i implies the less complexity of the DMPC, which achieves the decreased resource utilization burden. As a practitioner, it's recommended that the successive triggering instants satisfy $k_i^{d+1} \leqslant k_i^d + N$, i. e. the event triggering condition at time $k+1$ is formulated as

$$\Pi_i > \sigma_i \, \Xi_i \ or \ k+1=k_i^d+N \tag{2.24}$$

Moreover, the suitable value of parameter σ_i should be chosen such that the acceptableclosed-loop behavior and the resource utilization are achieved of the NCS.

Remark 2.3 To interpret the differences $\xi_i(k_i^d+l \mid k_i^d)$ and $\xi_j(k_i^d+l \mid k_i^d)$ on event-triggering condition (2.24), let k_i^d be the last triggering instant of subsystem $\mathcal{A}_i$. At time k_i^d+1, since $\xi_i(k_i^d+l \mid k_i^d)=\|x_i^*(k_i^d+l \mid k_i^d)-\hat{x}_i(k_i^d+l \mid k_i^d)\| \neq 0$, the event-triggering condition (2.24) is influenced by the differences $\xi_i(k_i^d+l \mid k_i^d)$ and $\xi_j(k_i^d+l \mid k_i^d)$ simultaneously. During the interval $k_i^d+1<k<k_i^{d+1}$, as $\xi_i(k+l \mid k)=\|\bar{x}_i(k+l \mid k)-\hat{x}_i(k+l \mid k)\|=\|\bar{x}_i(k+l \mid k-1)-\bar{x}_i(k+l \mid k-1)\|=0$, the event-triggering condition (2.24) only depends on the difference $\xi_j(k+l \mid k)$ from neighbor $\mathcal{A}_j$.

2.2.3 *Cooperative Event-Triggered DMPC Algorithm*

In the previous section, the basic formulation of DMPC problem and the triggering condition have been given. The proposed event-triggered DMPC can further be summarized as Algorithm 1.

Algorithm 1 Cooperative Event-triggered DMPC

1: **for** $k=0,1,2,\ldots$ **do**
2: **if** $k=0$ **then**
3 Let $\hat{x}_j(l \mid 0)=0 (l=0,\ldots,N-1)$. Based on $x_i(0)$, Problem 2.1 without constraint (2.9) is solved to obtain the optimal control sequence $u_i^*(0 \mid 0),\ldots,u_i^*(N-1 \mid 0)$, and then $u_i^*(0 \mid 0)$ is applied to $\mathcal{A}_i$. Subsystem $\mathcal{A}_i$ transmits $x_i^*(1 \mid 0)$ and $u_i^*(1 \mid 0),\ldots,u_i^*(N-1 \mid 0)$ to neighboring subsystem $\mathcal{A}_j$, and receives $x_j^*(1 \mid 0)$ and $u_j^*(1 \mid 0),\ldots,u_j^*(N-1 \mid 0)$ from subsystem $\mathcal{A}_j$.

---

else

  if the event-triggering condition (2.24) is satisfied **then**

  Update $k_i^d=k$. Problem 2.1 is solved based on $x_i(k_i^d)$ and the received predicted states $\overline{x}_j(k_i^d|k_i^d-1),\ldots,\overline{x}_j(k_i^d+N-1|k_i^d-1)$ to obtain the optimal control sequence $u_i^*(k_i^d|k_i^d),\ldots,u_i^*(k_i^d+N-1|k_i^d)$, and then $u_i^*(k_i^d|k_i^d)$ is applied to $\mathcal{A}_i$. Subsystem $\mathcal{A}_i$ transmits $x_i^*(k_i^d+1|k_i^d)$, and $u_i^*(k_i^d+1|k_i^d),\ldots,u_i^*(k_i^d+N-1|k_i^d)$ to neighboring subsystem $\mathcal{A}_j$, and receives $\overline{x}_j(k_i^d+1|k_i^d)$, and $\overline{u}_j(k_i^d+1|k_i^d),\ldots,\overline{u}_j(k_i^d+N-1|k_i^d)$ from subsystem $\mathcal{A}_j$.

else

  The control sequence (2.14) or (2.15) is considered and $u_i^*(k|k_i^d)$ is applied to $\mathcal{A}_i$. Subsystem $\mathcal{A}_i$ transmits $\overline{x}_i(k+1|k)$, and $\overline{u}_i(k+1|k),\ldots,\overline{u}_i(k+N-1|k)$ to neighboring subsystem $\mathcal{A}_j$, and receives $\overline{x}_j(k+1|k)$ and $\overline{u}_j(k+1|k),\ldots,\overline{u}_j(k+N-1|k)$ from subsystem $\mathcal{A}_j$

  end if

end if

end for

2.2.4 *Feasibility and Stability Analysis of the Overall Closed-Loop System*

The recursive feasibility and the global stability under event-triggered DMPC will be given in this section. Different from the time-triggered scheme, Problem 2.1 is executed only when the triggering condition is satisfied for each subsystems. The neighboring information is transferred in a asynchronous manner, which increase the difficulty of closed-loop properties analysis. In the next, we first prove that during two consecutive triggering instants, control sequence (2.14) or (2.15) is the feasible solution of Problem 2.1. The recursive feasibility of Algorithm 1 will be further concluded. Finally, the convergence of the closed-loop NCS system will be verified resorting to the input-to-state stability analysis techniques.

Theorem 2.1 *Consider the NCS (2.1) with Algorithm I implemented as the distributed controller, and suppose that for each subsystem $\mathcal{A}_i(i=1,\ldots,M)$. If there exists a feasible solution for Problem 2.1 at instant $k=k_i^d$, then Problem 2.1 is feasible at instant $k>k_i^d$.*

Proof Suppose that Problem 2.1 is solved at time k_i^d, let $u_i^*(k_i^d+l|k_i^d)$ and $x_i^*(k_i^d+l|k_i^d)$ be the optimal control sequence and the state trajectory. Next, we prove that control sequence (2.14) or (2.15) satisfies constraints (2.6)-(2.9) at instant $k>k_i^d$.

(i) $\bar{\boldsymbol{x}}_i(k+l|k)\in\boldsymbol{X}_i(l=1,\ldots,N-1)$.

For $k=k_i^d+1$, since $\boldsymbol{x}_i(k)=\boldsymbol{x}_i^*(k|k_i^k)$ and $\bar{\boldsymbol{u}}_i(k+l|k)=\boldsymbol{u}_i^*(k+l|k_i^d)$ $(l=0,\ldots,N-2)$, we can get $\bar{\boldsymbol{x}}_i(k+l|k)=\boldsymbol{x}_i^*(k+l|k_i^d)\in\boldsymbol{X}_i(l=1,\ldots,N-1)$. For $k_i^d+1<k<k_i^{d+1}$, since $\boldsymbol{x}_i(k)=\bar{\boldsymbol{x}}_i(k|k-1)$ and $\bar{\boldsymbol{u}}_i(k+l|k)=\bar{\boldsymbol{u}}_i(k+l|k-1)(l=0,\ldots,N-2)$, we can obtain $\bar{\boldsymbol{x}}_i(k+l|k)=\bar{\boldsymbol{x}}_i(k+l|k-1)\in\boldsymbol{X}_i(l=1,\ldots,N-1)$. Hence, for $k_i^d<k<k_i^{d+1}$, we have $\bar{\boldsymbol{x}}_i(k+l|k)\in\boldsymbol{X}_i(l=1,\ldots,N-1)$.

(ii) $\bar{\boldsymbol{x}}_i(k+N|k)\in\boldsymbol{X}_{fi}$.

For $k=k_i^d+1$, according to $\boldsymbol{x}_i^*(k_i^d+N|k_i^d)\in\boldsymbol{X}_{fi}$, we can obtain $\bar{\boldsymbol{x}}_i(k+N|k)=(\boldsymbol{A}_i+\boldsymbol{B}_i\boldsymbol{K}_i)\boldsymbol{x}_i^*(k_i^d+N|k_i^d)\in\boldsymbol{X}_{fi}$. For $k_i^d+1<k<k_i^{d+1}$, we have $\bar{\boldsymbol{x}}_i(k+N|k)=(\boldsymbol{A}_i+\boldsymbol{B}_i\boldsymbol{K}_i)\bar{\boldsymbol{x}}_i(k+N-1|k-1)\in\boldsymbol{X}_{fi}$. Therefore, we can get $\bar{\boldsymbol{x}}_i(k+N|k)\in\boldsymbol{X}_{fi}$ with $k_i^d<k<k_i^{d+1}$.

(iii) $\bar{\boldsymbol{u}}_i(k+l|k)\in\boldsymbol{U}_i(l=0,\ldots,N-1)$.

For $k=k_i^d+1$, it is easily shown that $\bar{\boldsymbol{u}}_i(k+l|k)=\boldsymbol{u}_i^*(k+l|k_i^d)\in\boldsymbol{U}_i(l=0,\ldots,N-2)$. Moreover, it follows from Assumption 2.1 that $\bar{\boldsymbol{u}}_i(k+N-1|k)=\boldsymbol{K}_i\boldsymbol{x}_i^*(k_i^d+N|k_i^d)\in\boldsymbol{U}_i$. For $k_i^d+1<k<k_i^{d+1}$, it follows from (2.15) that $\bar{\boldsymbol{u}}_i(k+l|k)=\bar{\boldsymbol{u}}_i(k+l|k-1)\in\boldsymbol{U}_i(l=0,\ldots,N-2)$. Furthermore, we can get $\bar{\boldsymbol{u}}_i(k+N-1|k)=\boldsymbol{K}_i\ \bar{\boldsymbol{x}}_i(k+N-1|k-1)\in\boldsymbol{U}_i$. Thus, for $k_i^d<k<k_i^{d+1}$, we have $\bar{\boldsymbol{u}}_i(k+l|k)\in\boldsymbol{U}_i(l=0,\ldots,N-1)$.

(iv) $\|\bar{\boldsymbol{x}}_i(k+l|k)-\hat{\boldsymbol{x}}_i(k+l|k)\|\leqslant\bar{\xi}_i(k)(l=1,\ldots,N)$.

For $k=k_i^d+1$, noting $\bar{\boldsymbol{x}}_i(k+l|k)=\boldsymbol{x}_i^*(k+l|k_i^d)$ and $\hat{\boldsymbol{x}}_i(k+l|k)=\boldsymbol{x}_i^*(k+l|k_i^d)$, we can get $\|\bar{\boldsymbol{x}}_i(k+l|k)-\hat{\boldsymbol{x}}_i(k+l|k)\|=0\leqslant\bar{\xi}_i(k)$. For $k_i^d+1<k<k_i^{d+1}$, since $\bar{\boldsymbol{x}}_i(k+l|k)=\bar{\boldsymbol{x}}_i(k+l|k-1)$ and $\hat{\boldsymbol{x}}_i(k+l|k)=\bar{\boldsymbol{x}}_i(k+l|k-1)$, we have $\|\bar{\boldsymbol{x}}_i(k+l|k)-\hat{\boldsymbol{x}}_i(k+l|k)\|=0\leqslant\bar{\xi}_i(k)$. Therefore, we can get $\|\bar{\boldsymbol{x}}_i(k+l|k)-\hat{\boldsymbol{x}}_i(k+l|k)\|\leqslant\bar{\xi}_i(k)$ with $k_i^d<k<k_i^{d+1}$.

□

According to Theorem 2.1, for any subsystem $\mathcal{A}_i(i=1,\ldots,M)$, Problem 2.1 is feasible at the next triggering instant k_i^{d+1} provided the feasibility of Problem 2.1 at instant k_i^d. Therefore, it can be concluded that Algorithm 1 for NCS (2.1) is recursively feasible. In the next, the basic definition and the lemma will be given according to the existing input-to-state stability concept.

Definition 2.1 The global $J(k)$ is an ISS Lyapunov function for global system (2.2), if for each $k \in N^*$, there exist $\mathcal{K}_\infty$ functions $\alpha_1, \alpha_2, \alpha_3$, and $\mathcal{K}$ function σ, such that

$$\alpha_1[\boldsymbol{x}(k)] \leqslant J(k) \leqslant \alpha_2[\boldsymbol{x}(k)],$$
$$J(k+1) - J(k) \leqslant -\alpha_3[\boldsymbol{x}(k)] + \sigma[\bar{\boldsymbol{\xi}}(k)]$$

with $\bar{\boldsymbol{\xi}}(k) = [\bar{\xi}_1(k), \ldots, \bar{\xi}_M(k)]^{\mathrm{T}}$ [10].

Lemma 2.1 *The global system (2.2) is ISS if it admits an ISS-Lyapunov function.*

Based on the above analysis tool, the sufficient condition for closed-loop convergence of global system (2.2) will be given in the following theorem.

Theorem 2.2 *The global system (2.2) is input-to-state stable, if at each time instant $k \in N^+$. there exists an upper bound of difference $\bar{\xi}_i(k)$ satisfying the following condition:*

$$\bar{\xi}_i(k) \leqslant \bar{\eta}_i(k) \tag{2.25}$$

where the scalar $\bar{\eta}_i(k)$ denotes the maximum difference between the estimated states $\hat{\boldsymbol{x}}_i(k+l \mid k)$ and $\hat{\boldsymbol{x}}_j(k+l \mid k)$ $(j \in \mathcal{N}_i)$ with $1 \leqslant l \leqslant N-1$, i.e., $\bar{\eta}_i(k) = \max\limits_{j \in \mathcal{N}_i} \max\limits_{1 \leqslant l \leqslant N-1} \{\eta_i(k+l \mid k)\}$.

Proof By Lemma 2.1, to prove the ISS of overall system (2.2), we need to show that $J(k)$ is an ISS-Lyapunov function for the overall system (2.2). In the following, we first consider the boundedness of $J(k)$. For all $\boldsymbol{x}_i(k) \in \boldsymbol{X}_i$, we have $J(k) \geqslant \sum_{i=1}^M \|\boldsymbol{x}_i(k)\|_{\boldsymbol{Q}_i}^2 \geqslant \sum_{i=1}^M \underline{\lambda}(\boldsymbol{Q}_i)\|\boldsymbol{x}_i(k)\|^2 \triangleq \alpha_1[\boldsymbol{x}(k)]$. Define $J(k, N) \triangleq J(k)$, and let $J(k, 0) = \sum_{i=1}^M \|\boldsymbol{x}_i(k)\|_{\boldsymbol{P}_i}^2$ with $\boldsymbol{x}_i(k) \in \boldsymbol{X}_{fi}$. By means of (2.13), the following inequality can be obtained:

$$J(k, N+1)$$
$$= \sum_{i=1}^{M}\Big\{\sum_{l=0}^{N}[\|\bar{\boldsymbol{x}}_i(k+l \mid k)\|_{\boldsymbol{Q}_i}^2 + \|\bar{\boldsymbol{u}}_i(k+l \mid k)\|_{\boldsymbol{R}_i}^2 +$$

$$\sum_{j\in\mathcal{N}_i}\|\bar{\boldsymbol{x}}_i(k+l\mid k)-\hat{\boldsymbol{x}}_j(k+l\mid k)\|_{\boldsymbol{Q}_{ij}}^2]$$
$$+\|\bar{\boldsymbol{x}}_i(k+N+1\mid k)\|_{\boldsymbol{P}_i}^2\}$$
$$=J(k,N)+\sum_{i=1}^{M}\{\|\bar{\boldsymbol{x}}_i(k+N+1\mid k)\|_{\boldsymbol{P}_i}^2-\|\bar{\boldsymbol{x}}_i(k+N\mid k)\|_{\boldsymbol{P}_i}^2+$$
$$\|\bar{\boldsymbol{x}}_i(k+N\mid k)\|_{\boldsymbol{Q}_i}^2$$
$$+\sum_{j\in\mathcal{N}_i}\|\bar{\boldsymbol{x}}_i(k+N\mid k)-\hat{\boldsymbol{x}}_j(k+N\mid k)\|_{\boldsymbol{Q}_{ij}}^2\}$$
$$\leqslant J(k,N)+\sum_{i=1}^{M}\|\bar{\boldsymbol{x}}_i(k+N\mid k)\|^2_{(\boldsymbol{A}_i+\boldsymbol{B}_i\boldsymbol{K}_i)^{\mathrm{T}}\boldsymbol{P}_i(\boldsymbol{A}_i+\boldsymbol{B}_i\boldsymbol{K}_i)-\boldsymbol{P}_i+\bar{\boldsymbol{Q}}_i+\sum_{j\in\mathcal{N}_i}2(\boldsymbol{Q}_{ij}+\boldsymbol{Q}_{ji})}$$
$$\leqslant J(k,N) \tag{2.26}$$

It can be further derived that $J(k,N)\leqslant\cdots\leqslant J(k,0)\leqslant\sum_{i=1}^{M}\bar{\lambda}(\boldsymbol{P}_i)\|\boldsymbol{x}_i(k)\|^2\triangleq\alpha_2[\boldsymbol{x}(k)]$ for all $\boldsymbol{x}_i(k)\in\boldsymbol{X}_{fi}$. Next, we consider the difference $\Delta J(k+1)$ in (2.20). Taking into account the condition (2.25) and interchanging the indices, we have

$$\sum_{i=1}^{M}\prod\nolimits_i\leqslant\sum_{i=1}^{M}\Big\{(N-1)\sum_{j\in\mathcal{N}_i}\bar{\lambda}(\boldsymbol{Q}_{ij})[4\bar{\eta}_i(k)\bar{\xi}_j(k)+\bar{\xi}_j^2(k)]\Big\}$$
$$=\sum_{i=1}^{M}\Big\{(N-1)\sum_{j\in\mathcal{N}_i}\bar{\lambda}(\boldsymbol{Q}_{ji})[4\bar{\eta}_j(k)\bar{\xi}_i(k)+\bar{\xi}_i^2(k)]\Big\}$$
$$\triangleq\sigma[\bar{\xi}(k)] \tag{2.27}$$

In addition, it is easily shown that

$$\sum_{i=1}^{M}\Xi_i\geqslant\sum_{i=1}^{M}\|\boldsymbol{x}_i(k)\|_{\boldsymbol{Q}_i}^2\geqslant\alpha_1[x(k)] \tag{2.28}$$

Thus, it follows from (2.27) and (2.28) that $J(k+1)-J(k)\leqslant-\alpha_1[\boldsymbol{x}(k)]+\sigma[\bar{\xi}(k)]$. Combining with Lemma 2.1, it can be concluded that the global system (2.2) is input-to-state stable, with $J(k)$ the candidate ISS Lyapunov function. □

Remark 2.4 The scalar $\bar{\xi}_i(k)$ essentially restricts the bias between the predicted state and the estimated state, for $\mathcal{A}_i$ at time instant k. Therefore, by including constraint (2.9) in the DMPC Problem 2.1 and applying the event-triggering condition (2.24), the algorithm feasibility and the input-to-state stability of the global system is ensured.

2.2.5 *Example*

In this section we consider the consensus problem of the position and speed variables, in a NCS composed of three agents. Each agent $\boldsymbol{A}_i$ $(i=1,2,3)$ is regarded as a subsystem, which is described as

$$\boldsymbol{x}_i(k+1)=\begin{bmatrix}1 & 0.1 & 0 & 0\\ 0 & 1 & 0 & 0\\ 0 & 0 & 1 & 0.1\\ 0 & 0 & 0 & 1\end{bmatrix}\boldsymbol{x}_i(k)+\begin{bmatrix}0.05 & 0\\ 1 & 0\\ 0 & 0.05\\ 0 & 1\end{bmatrix}\boldsymbol{u}_i(k). \quad (2.29)$$

in which the state vector is defined as $\boldsymbol{x}_i=[x_{i1},x_{i2},x_{i3},x_{i4}]^{\mathrm{T}}$, x_{i1} is the horizontal position s_i^x, x_{i3} is the vertical position s_i^y, x_{i2} and x_{i4} are the horizontal velocity v_i^x and vertical velocity v_i^y, respectively; the input vector is defined as $\boldsymbol{u}_i=[u_{i1},u_{i2}]^{\mathrm{T}}$, u_{i1} is the horizontal force f_i^x and u_{i2} is the vertical force f_i^y.

The constraint on control input is $\|\boldsymbol{u}_i(k)\|_\infty\leqslant 0.5$. The parameters in Problem 2.1 are chosen as $N=10$, $\boldsymbol{Q}_i=I_{4\times 4}$, $\boldsymbol{R}_i=I_{2\times 2}$, $\mathcal{N}_1=\{2,3\}$, $\mathcal{N}_2=\{1,3\}$, $\mathcal{N}_3=\{1,2\}$, $\boldsymbol{Q}_{ij}=[1,0,0,0;0,0,0,0;0,0,1,0;0,0,0,0]$. The triggering parameters for three agents are set as $\sigma_1=0.9$, $\sigma_2=0.6$ and $\sigma_3=0.3$. The initial conditions are given by $\boldsymbol{x}_1(0)=[0.4,0,-0.5,0]^{\mathrm{T}}$, $\boldsymbol{x}_2(0)=[0.3,0,0.6,0]^{\mathrm{T}}$ and $\boldsymbol{x}_3(0)=[-0.5,0,-0.4,0]^{\mathrm{T}}$.

The closed-loop horizontal velocities and vertical velocities of three agents under Algorithm 1 are depicted in Fig. 2.2; the horizontal forces and vertical forces are plotted in Fig. 2.3. From Fig. 2.2 and Fig. 2.3. It can be observed that the system achieves consensus w.r.t. velocity variable. The control input satisfy the constraints. The triggering instants for three agents are presented in Fig. 2.4. The results indicate that for each agent, Problem 2.1 is solved in an aperiodic manner, which reduces communicational and computational costs. To further clarify the impacts of triggering parameter σ_i on the resource utilization, the trajectories of three agents with different triggering parameter σ_i are presented in Fig. 2.5, and the whole problem-solving times n and the percent of problem-solving times p are listed in Table 2.1. It can be concluded that enlarging the triggering parameter σ_i causes a decreasing of resource consumption, which is consistent with the previous analysis.

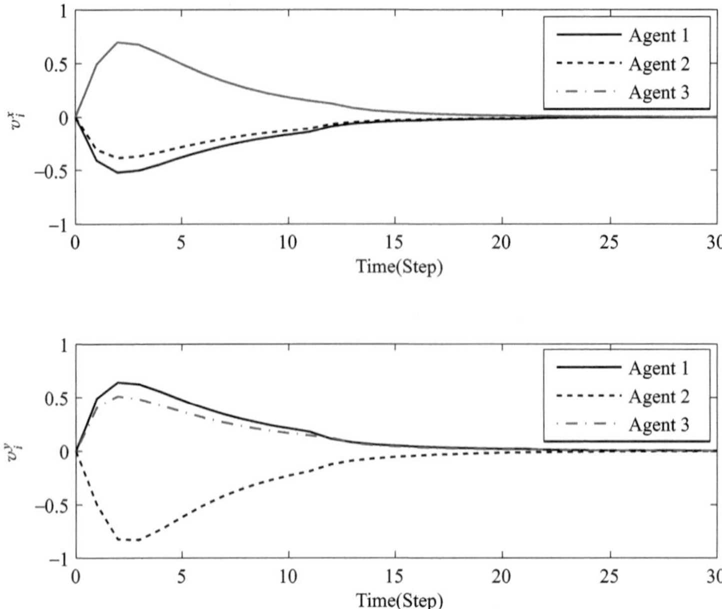

Fig. 2. 2 Horizontal velocities and vertical velocities of three agents

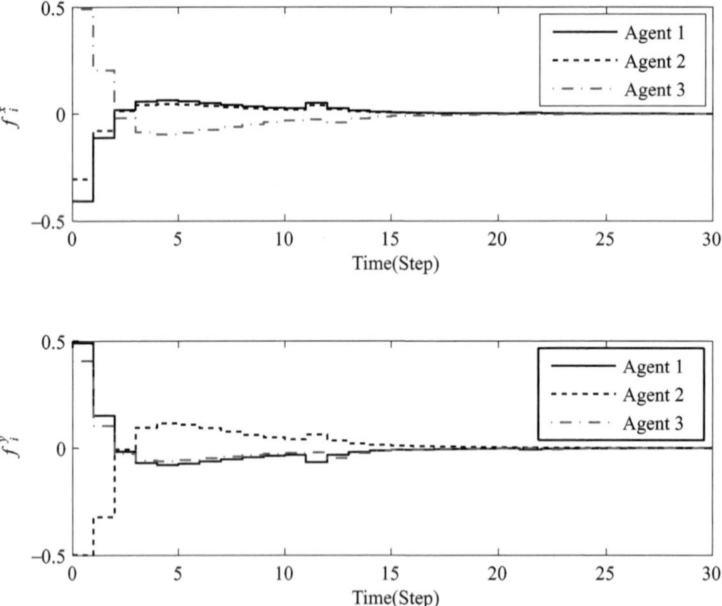

Fig. 2. 3 Horizontal forces and vertical forces of three agents

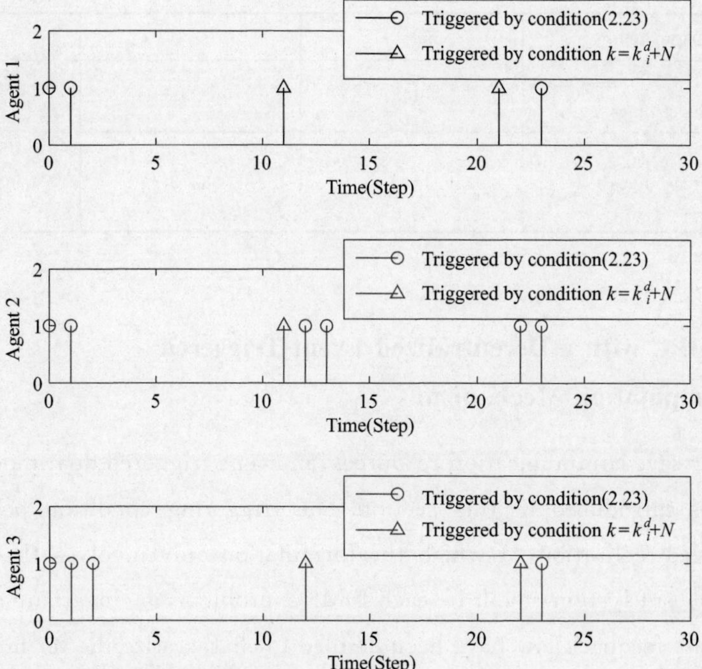

Fig. 2.4 Triggering instants for three agents

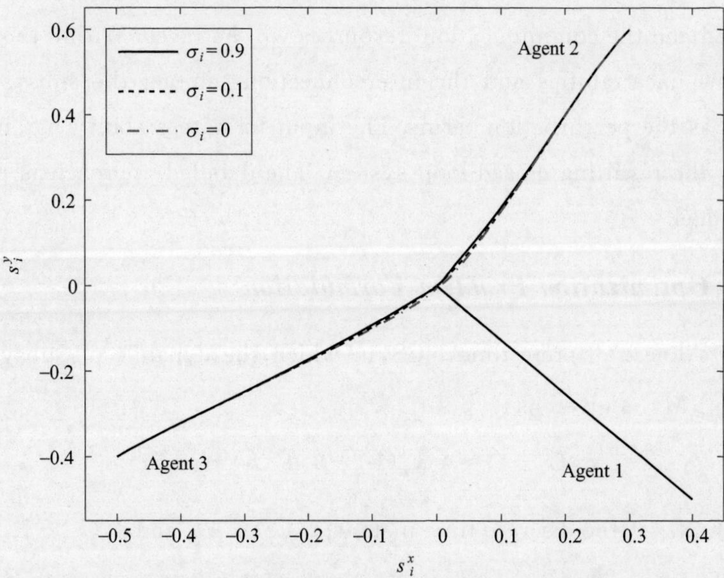

Fig. 2.5 Trajectories of three agents with different triggering parameter σ_i

Table 2. 1 Results with different triggering parameter σ_i

Triggering parameter σ_i	0. 9	0. 1	0
Whole problem-solving times n	14	29	90
Percent of problem-solving times p(%)	15. 6	32. 2	100

2. 3 DMPC with a Decentralized Event-Triggered Computation Mechanism

To further save communication resources, an event-triggered dual-mode DMPC strategy is introduced in this section. The triggering condition possesses a decentralized realization, of which the formulation only involves the information of the subsystem itself. In each DMPC problem, an invariant set and a related state feedback law have been defined to characterize the terminal condition. The local state feedback control will be applied, once the state enters the invariant set. In the communication level, if the states enter the invariant sets of the subsystem and its neighbors respectively, no information will be exchanged and the communication resources will be saved. Within the terminal region, the uncertainties and the interconnections among the subsystems are regarded as the perturbation terms. The input-to-state stability has been analyzed for the resulting closed-loop system. The detailed approach is presented in the sequel.

2. 3. 1 Optimization Problem Formulation

Consider a linear discrete-time NCS, in which the dynamic of subsystem $\mathcal{A}_i$ ($i=1,\ldots,M$) is given as

$$\boldsymbol{x}_i(k+1)=\boldsymbol{A}_i\boldsymbol{x}_i(k)+\boldsymbol{B}_i\boldsymbol{u}_i(k)+\boldsymbol{w}_i(k) \tag{2.30}$$

where $k\in N^+$ is the discrete time instant; $\boldsymbol{x}_i(k)\in\mathbb{R}^{n_i}$ and $\boldsymbol{u}_i(k)\in\mathbb{R}^{m_i}$ are the state and input vectors respectively; $\boldsymbol{w}_i(k)\in\mathbb{R}^{n_i}$ is the perturbation term; $\boldsymbol{A}_i\in\mathbb{R}^{n_i\times n_i}$ and $\boldsymbol{B}_i\in\mathbb{R}^{n_i\times m_i}$ are the constant matrices. For every subsystem

$\mathcal{A}$,$(\boldsymbol{A}_i, \boldsymbol{B}_i)$ is assumed to be stabilizable. The state and input constraints are formulated as

$$\boldsymbol{x}_i(k) \in \boldsymbol{X}_i = \{\boldsymbol{x}_i \in \mathbb{R}^{n_i} : \|\boldsymbol{x}_i\| \leqslant \overline{\boldsymbol{x}}_i\} \tag{2.31}$$

$$\boldsymbol{u}_i(k) \in \boldsymbol{U}_i = \{\boldsymbol{u}_i \in \mathbb{R}^{m_i} : \|\boldsymbol{u}_i\| \leqslant \overline{\boldsymbol{u}}_i\} \tag{2.32}$$

where $\boldsymbol{X}_i$ and $\boldsymbol{U}_i$ are compact sets that contain the origin. The perturbation term $\boldsymbol{w}_i(k)$ is assumed to be bounded, i. e. $\|\boldsymbol{w}_i\| \leqslant \overline{\boldsymbol{w}}_i$. The nominal subsystem related to $\mathcal{A}_i$ is defined as :

$$\boldsymbol{x}_i(k+1) = \boldsymbol{A}_i \boldsymbol{x}_i(k) + \boldsymbol{B}_i \boldsymbol{u}_i(k) \tag{2.33}$$

The control structure of the NCS is referred to as Fig. 2. 1. At every sampling instant, the trigger for every subsystem determines whether to execute the sub DMPC controller. Different from the approach in Sect. 2. 2, the triggering condition is parameterized at the local state only. Utilizing this mechanism reduces the communicational burden to some extent; however, the estimation error caused by system uncertainty and the subsystems' interaction should be carefully treated in the DMPC formulation, such that the expected closed-loop property can be achieved. Let k_i^d $(d \in \mathbb{N})$ be the d -th triggering instant of the DMPC for subsystem $\mathcal{A}_i (i=1, \ldots, M)$; and initialize k_i^d with $k_i^0=0$. The DMPC solves following FHOCP repeatedly w. r. t. the triggering instants k_i^d $(d \in \mathbb{N})$.

Problem 2. 2 Let $\boldsymbol{u}_i(k_i^d + l \mid k_i^d)(l=0, \ldots, N-1)$ be the input sequence to be optimized; let $\boldsymbol{x}_i(k_i^d + l \mid k_i^d)(l=0, \ldots, N)$ be the predicted state sequence emanating from the measurement at instant k_i^d. The DMPC optimization problem to be solved at time k_i^d is given as

$$\min_{\boldsymbol{u}_i(k_i^d + l \mid k_i^d), 0 \leqslant l \leqslant N-1} J_i(k_i^d) \tag{2.34}$$

subject to:

$$\boldsymbol{x}_i(k_i^d + l + 1 \mid k_i^d) = \boldsymbol{A}_i \boldsymbol{x}_i(k_i^d + l \mid k_i^d) + \boldsymbol{B}_i \boldsymbol{u}_i(k_i^d + l \mid k_i^d) \tag{2.35}$$

$$\boldsymbol{x}_i(k_i^d + l \mid k_i^d) \in \boldsymbol{X}_i^l = \boldsymbol{X}_i \sim \boldsymbol{W}_i^l, l=1, \ldots, N-1 \tag{2.36}$$

$$\boldsymbol{u}_i(k_i^d + l \mid k_i^d) \in \boldsymbol{U}_i, l=0, \ldots, N-1 \tag{2.37}$$

$$\boldsymbol{x}_i(k_i^d + N \mid k_i^d) \in \Phi_i \tag{2.38}$$

$$J_i^E(k_i^d) \leqslant \varphi_i(k_i^d) \tag{2.39}$$

with the cost index defined as $J_i(k_i^d)=J_i^E(k_i^d)+J_i^F(k_i^d)$, and

$$J_i^E(k_i^d) \triangleq \sum_{l=0}^{N-1}[\|\boldsymbol{x}_i(k_i^d+l|k_i^d)\|_{\boldsymbol{Q}_i}^2+\|\boldsymbol{u}_i(k_i^d+l|k_i^d)\|_{\boldsymbol{R}_i}^2]+\|\boldsymbol{x}_i(k_i^d+N|k_i^d)\|_{\boldsymbol{P}_i}^2,$$

$$J_i^F(k_i^d) \triangleq \sum_{l=0}^{N-1}\sum_{j\in\mathcal{N}_i}\|\boldsymbol{x}_i(k_i^d+l|k_i^d)-\boldsymbol{x}_j^a(k_i^d+l|k_i^d)\|_{\boldsymbol{Q}_{ij}}^2,$$

where $J_i^E(k_i^d)$ is the local regulation term and $J_i^F(k_i^d)$ is the cooperative term ; N is the prediction horizon; $\boldsymbol{Q}_i$, $\boldsymbol{R}_i$, $\boldsymbol{P}_i$ and $\boldsymbol{Q}_{ij}$ are the positive definite weighting matrices; $\mathcal{N}_i$ is the set of indexes of $\mathcal{A}_i$'s neighbors ; $\boldsymbol{x}_j^a(k+l|k)$ is the estimated state of subsystem $\mathcal{A}_j$, based on the measurement of $\mathcal{A}_i$ at instant k and information of $\mathcal{A}_j$ received at time instant $k-1$. According to the last triggering instant of $\mathcal{A}_i$, the $x_j^a(k+l|k)$ is specifically defined as follows.

(1) For the case with triggering instant of $\mathcal{A}_j$ being k_i^d-1, there is

$$\boldsymbol{x}_j^a(k_i^d+l|k_i^d)=\boldsymbol{x}_j^*(k_i^d+l|k_i^d-1), l=0,\ldots,N-1$$

where $\boldsymbol{x}_j^*(k+l|k-1)$ denotes the optimal state trajectory of Problem 2.2 at time k_i^d-1.

(2) For the case with triggering instant of $\mathcal{A}_j$ earlier than k_i^d-1, there is

$$\boldsymbol{x}_j^a(k_i^d+l|k_i^d)=\overline{\boldsymbol{x}}_j(k_i^d+l|k_i^d-1), l=0,\ldots,N-1$$

where $\overline{\boldsymbol{x}}_j(k_i^d+l|k_i^d-1)$ is the feasible state trajectory at time k_i^d-1. The construction of feasible state/input trajectories follows the approach in Sect. 2.2. The shifted control input is derived by the optimized results at the last triggering instant and the local state feedback control law K_i. However, K_i is defined in a different form in order to cope with the uncertainty, which will be introduced in detail in the next subsection.

Specifically, the predicted states in constraint (2.36) are restricted by the Pontryagin difference of $\boldsymbol{X}_i$ and $\boldsymbol{W}_i^l$, with

$$\boldsymbol{W}_i^l=\left\{\boldsymbol{w}_i\in\mathbb{R}^{n_i}:\|\boldsymbol{w}_i\|\leqslant\frac{1-\|\boldsymbol{A}_i\|^l}{1-\|\boldsymbol{A}_i\|}\overline{\boldsymbol{w}}_i\right\}$$

This implies that $\boldsymbol{x}_i(k_i^d+l|k_i^d)+\boldsymbol{w}_i(k_i^d)\in\boldsymbol{X}_i$, for $\forall\, \boldsymbol{w}_i(k_i^d)\in\boldsymbol{W}_i^l$. The terminal state is restricted by constraint (2.38), in which $\Phi_i=\{\boldsymbol{x}_i\in\mathbb{R}^{n_i}:\|\boldsymbol{x}_i\|_{\boldsymbol{P}_i}^2\leqslant\overline{\varepsilon}_i\}$.

Constraint (2.39) regulates the upper bound of the local cost $J_i^E(k_i^d)$, which is crucial to achieve closed-loop convergence. $\varphi_i(k_i^d)$ is a scalar that varies over time, and its specific definition will be given in the subsequent section.

2.3.2 *Decentralized Event-Triggering Condition*

The event-triggering condition is given based on the ISS theory and the closed-loop convergence can be guaranteed. An event-triggered dual-mode DMPC algorithm is further given, which saves the computational and communicational resources for large-scale NCSs with bounded disturbances.

As a primary step, we first characterize an invariant set and the corresponding state feedback law in the following lemma.

Lemma 2.2 *For the nominal subsystem (2.33) with stabilizable pair* $(\boldsymbol{A}_i, \boldsymbol{B}_i)$, *let* $\boldsymbol{Q}_i$ *and* $\boldsymbol{R}_i$ *be any positive definite symmetric matrices, there exist a positive definite symmetric matrix* $\boldsymbol{P}_i$, *a state feedback control law* $\boldsymbol{K}_i$, *constants* $\varepsilon_i > 0$ *and* $0 < \bar{\varepsilon}_i < \varepsilon_i$ *such that*: 1) $\forall \boldsymbol{x}_i \in \phi_i$, $\boldsymbol{K}_i\boldsymbol{x}_i \in \boldsymbol{U}_i$ *and* $\|(\boldsymbol{A}_i+\boldsymbol{B}_i\boldsymbol{K}_i)\boldsymbol{x}_i\|_{\boldsymbol{P}_i}^2 - \|\boldsymbol{x}_i\|_{\boldsymbol{P}_i}^2 \leqslant -\|\boldsymbol{x}_i\|_{\overline{\boldsymbol{Q}}_i}^2$, *where* $\phi_i = \{\boldsymbol{x}_i \in \mathbb{R}^{n_i} : \|\boldsymbol{x}_i\|_{\boldsymbol{P}_i}^2 \leqslant \varepsilon_i\}$, $\overline{\boldsymbol{Q}}_i = \boldsymbol{Q}_i + \boldsymbol{K}_i^{\mathrm{T}}\boldsymbol{R}_i\boldsymbol{K}_i$; 2) $\forall \boldsymbol{x}_i \in \phi_i$, $(\boldsymbol{A}_i+\boldsymbol{B}_i\boldsymbol{K}_i)\boldsymbol{x}_i \in \phi_i$.

Proof The results can be derived from [12]. □

In the next, we construct the feasible control sequence between two consecutive triggering instant k_i^d and k_i^{d+1}, w.r.t. subsystem $\mathcal{A}_i$, Denote the optimal control sequence of Problem 2.2 at time instant k_i^d as $\boldsymbol{u}_i^*(k_i^d+l \mid k_i^d)$; denote the related optimal state trajectory $\boldsymbol{x}_i^*(k_i^d+l \mid k_i^d)$ and the optimal cost as $J_i^*(k_i^d)$, with $J_i^*(k_i^d) = J_i^{E*}(k_i^d) + J_i^{F*}(k_i^d)$. For time interval $k \in [k_i^d, k_i^{d+1})$. the feasible control sequence $\bar{\boldsymbol{u}}_i(k+l \mid k)$ is defined as follows.

For $k = k_i^d + 1$,

$$\bar{\boldsymbol{u}}_i(k+l \mid k) = \begin{cases} \boldsymbol{u}_i^*(k+l \mid k_i^d), l=0,\ldots,N-2 \\ \boldsymbol{K}_i\bar{\boldsymbol{x}}_i(k+l \mid k), l=N-1 \end{cases} \tag{2.40}$$

For $k_i^d + 1 < k < k_i^{d+1}$,

$$\bar{\boldsymbol{u}}_i(k+l|k)=\begin{cases}\bar{\boldsymbol{u}}_i(k+l|k-1),l=0,\ldots,N-2\\ \boldsymbol{K}_i\bar{\boldsymbol{x}}_i(k+l|k),l=N-1\end{cases} \tag{2.41}$$

Denote the values of $J_i^E(k)$ under the feasible control sequence $\bar{\boldsymbol{u}}_i(k+l|k)$ and $\bar{\boldsymbol{u}}_i(k+l|k-1)$ as $\bar{J}_i^E(k)$ and $\bar{J}_i^E(k-1)$, respectively. Let $\bar{J}_i^E(k_i^d)=J_i^{E*}(k_i^d)$. The difference between $\bar{J}_i^E(k)$ and $\bar{J}_i^E(k-1)$ is given by

$$\begin{aligned}
&\Delta J_i^E(k)\\
&=\bar{J}_i^E(k)-\bar{J}_i^E(k-1)\\
&=\sum_{l=0}^{N-1}[\|\bar{\boldsymbol{x}}_i(k+l|k)\|_{\boldsymbol{Q}_i}^2+\|\bar{\boldsymbol{u}}_i(k+l|k)\|_{\boldsymbol{R}_i}^2]+\|\bar{\boldsymbol{x}}_i(k+N|k)\|_{\boldsymbol{P}_i}^2-\\
&\quad\sum_{l=-1}^{N-2}[\|\bar{\boldsymbol{x}}_i(k+l|k-1)\|_{\boldsymbol{Q}_i}^2+\|\bar{\boldsymbol{u}}_i(k+l|k-1)\|_{\boldsymbol{R}_i}^2]-\|\bar{\boldsymbol{x}}_i(k+N-1|k-1)\|_{\boldsymbol{P}_i}^2\\
&=\sum_{l=0}^{N-2}[\|\bar{\boldsymbol{x}}_i(k+l|k)\|_{\boldsymbol{Q}_i}^2+\|\bar{\boldsymbol{u}}_i(k+l|k)\|_{\boldsymbol{R}_i}^2-\|\bar{\boldsymbol{x}}_i(k+l|k-1)\|_{\boldsymbol{Q}_i}^2\\
&\quad-\|\bar{\boldsymbol{u}}_i(k+l|k-1)\|_{\boldsymbol{R}_i}^2]\\
&\quad+\|\bar{\boldsymbol{x}}_i(k+N-1|k)\|_{\boldsymbol{Q}_i}^2-\|\boldsymbol{x}_i(k-1)\|_{\boldsymbol{Q}_i}^2-\|\boldsymbol{u}_i^*(k-1|k_i^d)\|_{\boldsymbol{R}_i}^2\\
&\quad+\|(\boldsymbol{A}_i+\boldsymbol{B}_i\boldsymbol{K}_i)\bar{\boldsymbol{x}}_i(k+N-1|k)\|_{\boldsymbol{P}_i}^2-\|\bar{\boldsymbol{x}}_i(k+N-1|k-1)\|_{\boldsymbol{P}_i}^2\\
&\quad+\|\bar{\boldsymbol{x}}_i(k+N-1|k)\|_{\boldsymbol{P}_i}^2-\|\bar{\boldsymbol{x}}_i(k+N-1|k)\|_{\boldsymbol{P}_i}^2
\end{aligned} \tag{2.42}$$

From Lemma 2.2, we have

$$\|(\boldsymbol{A}_i+\boldsymbol{B}_i\boldsymbol{K}_i)\bar{\boldsymbol{x}}_i(k+N-1|k)\|_{\boldsymbol{P}_i}^2-\|\bar{\boldsymbol{x}}_i(k+N-1|k)\|_{\boldsymbol{P}_i}^2+\|\bar{\boldsymbol{x}}_i(k+N-1|k)\|_{\boldsymbol{Q}_i}^2\leqslant 0$$

Since $\bar{\boldsymbol{u}}_i(k+l|k)=\bar{\boldsymbol{u}}_i(k+l|k-1)(l=0,\ldots,N-2)$, there is

$$\begin{aligned}
&\|\bar{\boldsymbol{x}}_i(k+l|k)-\bar{\boldsymbol{x}}_i(k+l|k-1)\|\\
&=\left\|\boldsymbol{A}_i^l\boldsymbol{x}_i(k)+\sum_{p=1}^{l}\boldsymbol{A}_i^{l-p}\boldsymbol{B}_i\bar{\boldsymbol{u}}_i(k+p-1|k)-\boldsymbol{A}_i^l\bar{\boldsymbol{x}}_i(k|k-1)\right.\\
&\quad\left.-\sum_{p=1}^{l}\boldsymbol{A}_i^{l-p}\boldsymbol{B}_i\bar{\boldsymbol{u}}_i(k+p-1|k-1)\right\|\\
&=\|\boldsymbol{A}_i^l[\boldsymbol{A}_i\boldsymbol{x}_i(k-1)+\boldsymbol{B}_i\bar{\boldsymbol{u}}_i(k-1|k-1)+\boldsymbol{w}_i(k-1)-\boldsymbol{A}_i\boldsymbol{x}_i(k-1)\\
&\quad-\boldsymbol{B}_i\bar{\boldsymbol{u}}_i(k-1|k-1)]\|\\
&\leqslant\|\boldsymbol{A}_i\|^l\bar{\boldsymbol{w}}_i
\end{aligned} \tag{2.43}$$

By means of (2.31) and(2.43), there is

$$
\begin{aligned}
&\|\bar{\boldsymbol{x}}_i(k+l\,|\,k)\|_{\boldsymbol{Q}_i}^2-\|\bar{\boldsymbol{x}}_i(k+l\,|\,k-1)\|_{\boldsymbol{Q}_i}^2\\
\leqslant{}&2\bar{\lambda}(\boldsymbol{Q}_i)\|\bar{\boldsymbol{x}}_i(k+l\,|\,k-1)\|\cdot\|\bar{\boldsymbol{x}}_i(k+l\,|\,k)-\bar{\boldsymbol{x}}_i(k+l\,|\,k-1)\|\\
&+\bar{\lambda}(\boldsymbol{Q}_i)\|\bar{\boldsymbol{x}}_i(k+l\,|\,k)-\bar{\boldsymbol{x}}_i(k+l\,|\,k-1)\|^2\\
\leqslant{}&2\bar{\lambda}(\boldsymbol{Q}_i)\bar{x}_i\|\boldsymbol{A}_i\|^l\,\bar{w}_i+\bar{\lambda}(\boldsymbol{Q}_i)\|\boldsymbol{A}_i\|^{2l}\,\bar{w}_i^2
\end{aligned}
\tag{2.44}
$$

Similarly, there is

$$
\begin{aligned}
&\|\bar{\boldsymbol{x}}_i(k+N-1\,|\,k)\|_{\boldsymbol{P}_i}^2-\|\bar{\boldsymbol{x}}_i(k+N-1\,|\,k-1)\|_{\boldsymbol{P}_i}^2\\
\leqslant{}&2\bar{\lambda}(\boldsymbol{P}_i)\|\bar{\boldsymbol{x}}_i(k+N-1\,|\,k-1)\|\cdot\|\bar{\boldsymbol{x}}_i(k+N-1\,|\,k)-\bar{\boldsymbol{x}}_i(k+N-1\,|\,k-1)\|\\
&+\bar{\lambda}(\boldsymbol{P}_i)\|\bar{\boldsymbol{x}}_i(k+N-1\,|\,k)-\bar{\boldsymbol{x}}_i(k+N-1\,|\,k-1)\|^2\\
\leqslant{}&2\bar{\lambda}(\boldsymbol{P}_i)\bar{x}_i\|\boldsymbol{A}_i\|^{N-1}\,\bar{w}_i+\bar{\lambda}(\boldsymbol{P}_i)\|\boldsymbol{A}_i\|^{2(N-1)}\bar{w}_i^2
\end{aligned}
\tag{2.45}
$$

Summing (2.44) from 0 to $N-2$, it derives

$$
\begin{aligned}
&\sum_{l=0}^{N-2}\left[\|\bar{\boldsymbol{x}}_i(k+l\,|\,k)\|_{\boldsymbol{Q}_i}^2-\|\bar{\boldsymbol{x}}_i(k+l\,|\,k-1)\|_{\boldsymbol{Q}_i}^2\right]\\
\leqslant{}&\frac{2\bar{\lambda}(\boldsymbol{Q}_i)\bar{x}_i(1-\|\boldsymbol{A}_i\|^{N-1})\bar{w}_i}{1-\|\boldsymbol{A}_i\|}+\frac{\bar{\lambda}(\boldsymbol{Q}_i)(1-\|\boldsymbol{A}_i\|^{2(N-1)})\bar{w}_i^2}{1-\|\boldsymbol{A}_i\|^2}
\end{aligned}
\tag{2.46}
$$

Substituting (2.45)-(2.46) into (2.42), it follows that

$$
\bar{J}_i^E(k)\leqslant\bar{J}_i^E(k-1)-\|\boldsymbol{x}_i(k-1)\|_{\boldsymbol{Q}_i}^2-\|\boldsymbol{u}_i^*(k-1\,|\,k_i^d)\|_{\boldsymbol{R}_i}^2+\Theta_i(\bar{w}_i)
\tag{2.47}
$$

where

$$
\begin{aligned}
\Theta_i(\bar{w}_i)={}&2\bar{x}_i\bar{w}_i\left[\bar{\lambda}(\boldsymbol{P}_i)\|\boldsymbol{A}_i\|^{N-1}+\frac{\bar{\lambda}(\boldsymbol{Q}_i)(1-\|\boldsymbol{A}_i\|^{N-1})}{1-\|\boldsymbol{A}_i\|}\right]\\
&+\bar{w}_i^2\left[\bar{\lambda}(\boldsymbol{P}_i)\|\boldsymbol{A}_i\|^{2(N-1)}+\frac{\bar{\lambda}(\boldsymbol{Q}_i)(1-\|\boldsymbol{A}_i\|^{2(N-1)})}{1-\|\boldsymbol{A}_i\|^2}\right]
\end{aligned}
$$

Let

$$
\Theta_i(\bar{w}_i)\leqslant\sigma_i\left[\|\boldsymbol{x}_i(k-1)\|_{\boldsymbol{Q}_i}^2+\|\boldsymbol{u}_i^*(k-1\,|\,k_i^d)\|_{\boldsymbol{R}_i}^2\right]
$$

with $0<\sigma_i<1$. Thus, there is

$$
\Delta J_i^E(k)\leqslant(\sigma_i-1)\left[\|\boldsymbol{x}_i(k-1)\|_{\boldsymbol{Q}_i}^2+\|\boldsymbol{u}_i^*(k-1\,|\,k_i^d)\|_{\boldsymbol{R}_i}^2\right]<0
$$

To maintain the monotonically decreasing property of this function, we define the event-triggering condition of subsystem $\mathcal{A}_i$ at time k as

$$\Theta_i(\overline{w}_i) > \sigma_i [\|x_i(k-1)\|_{Q_i}^2 + \|u_i^*(k-1|k_i^d)\|_{R_i}^2]$$

To guarantee performance, it's required that the successive triggering instants satisfy $k_i^{d+1} \leqslant k_i^d + N$, i. e. the event triggering condition for subsystem $\mathcal{A}_i$ at time k is formulated as

$$\Theta_i(\overline{w}_i) > \sigma_i [\|x_i(k-1)\|_{Q_i}^2 + \|u_i^*(k-1|k_i^d)\|_{R_i}^2] \\ or\ k = k_i^d + N \tag{2.48}$$

Remark 2.5 It should be noticed that, the value of $\|x_i(k-1)\|_{Q_i}^2 + \|u_i^*(k-1|k_i^d)\|_{R_i}^2$ will become increasingly smaller upon the convergence of the system. This essentially results in the persistent satisfaction of condition (2.48). To avoid the excessive consumption of computational and communicational resources, a dual-mode control strategy is adopted. When subsystem $\mathcal{A}_i$ enters its invariant set ϕ_i, the state feedback control law K_i is applied instead of solving the Problem 2.2. At this stage, if the states of subsystem $\mathcal{A}_i$'s neighbors enter their invariant sets as well, no information will be exchanged.

2.3.3 *Event-Triggered Dual-Mode DMPC Algorithm*

On the basis of Problem 2.2 and the triggering condition (2.48), we propose the event-triggered dual-mode DMPC algorithm as follows.

2.3.4 *Feasibility and Stability Analysis of the Overall Closed-Loop System*

In this section, the sufficient conditions for ensuring the feasibility and closed-loop stability are derived for the proposed event-triggered dual-mode DMPC algorithm. To proceed, the following lemma is given.

Algorithm 2 Event-triggered Dual-mode DMPC

for $k=0,1,2,\ldots$ **do**
  if $k=0$ **then**
    Let $x_j^a(l|0)=0(l=0,\ldots,N-1)$ and $\varphi_i(0)=\infty$. Based on $x_i(0)$. Problem 2.2 is solved to obtain the optimal control sequence $u_i^*(0|0),\ldots,u_i^*(N-1|0)$, and then $u_i^*(0|0)$ is applied to $\mathcal{A}_i$. Subsystem $\mathcal{A}_i$ transmits $x_i^*(1|0),\ldots,x_i^*(N|0)$ to neighboring subsystem $\mathcal{A}_j$, and receives $x_j^*(1|0),\ldots,x_j^*(N|0)$ from $\mathcal{A}_j$.
  else
    if $x_i(k)\in\phi_i$ **then**
      Apply $u_i(k)=K_i x_i(k)$ to subsystem $\mathcal{A}_i$. $\mathcal{A}_i$ transmits the associated state trajectory $(A_i+B_iK_i)x_i(k),\ldots,(A_i+B_iK_i)^N x_i(k)$ to subsystem $\mathcal{A}_j$ when $x_j(k)\notin\phi_j$.

---

else

if the event-tiggering condition (2.48) is satisfied **then**

Update $k_i^d = k$. Problem 2.2 is solved to obtain the optimal control sequence $\boldsymbol{u}_i^*(k_i^d \mid k_i^d), \ldots, \boldsymbol{u}_i^*(k_i^d + N - 1 \mid k_i^d)$, and then $\boldsymbol{u}_i^*(k_i^d \mid k_i^d)$ is applied to $\mathcal{A}_i$. Subsystem $\mathcal{A}_i$ transmits $\boldsymbol{x}_i^*(k_i^d + 1 \mid k_i^d), \ldots, \boldsymbol{x}_i^*(k_i^d + N \mid k_i^d)$ to neighboring subsystem $\mathcal{A}_j$, and receives $\overline{\boldsymbol{x}}_j(k_i^d + 1 \mid k_i^d), \ldots, \overline{\boldsymbol{x}}_j(k_i^d + N \mid k_i^d)$ from subsystem $\mathcal{A}_j$.

else

The feasible control sequence (2.40)or (2.41) is constructed and $\overline{\boldsymbol{u}}_i(k \mid k_i^d)$ is applied to $\mathcal{A}_i$. Subsystem $\mathcal{A}_i$ transmits $\overline{\boldsymbol{x}}_i(k+1 \mid k), \ldots, \overline{\boldsymbol{x}}_i(k+N \mid k)$ to neighboring subsystem $\mathcal{A}_j$, and receives $\overline{\boldsymbol{x}}_j(k+1 \mid k), \ldots, \overline{\boldsymbol{x}}_j(k+N \mid k)$ from subsystem $\mathcal{A}_j$.

end if

end if

end if

end for

Lemma 2.3 *Define* $\boldsymbol{X}_i^{l+1} = \boldsymbol{X}_i \sim \boldsymbol{W}_i^{l+1}$ $(l = 1, \ldots, N-1)$, *if* $\boldsymbol{x} \in \boldsymbol{X}_i^{l+1}$ *and* $\boldsymbol{y} \in \mathbb{R}^{n_i}$ *satisfy*

$$\| \boldsymbol{x} - \boldsymbol{y} \| \leqslant \| \boldsymbol{A}_i \|^l \overline{\boldsymbol{w}}_i \tag{2.49}$$

then $\boldsymbol{y} \in \boldsymbol{X}_i^l$.

Proof Let $\boldsymbol{z} = \boldsymbol{y} - \boldsymbol{x} + \boldsymbol{v}$, where $\boldsymbol{v} \in \boldsymbol{W}_i^l$. According to (2.49), there is

$$\begin{aligned} \| \boldsymbol{z} \| &= \| \boldsymbol{y} - \boldsymbol{x} + \boldsymbol{v} \| \\ &\leqslant \| \boldsymbol{x} - \boldsymbol{y} \| + \| \boldsymbol{v} \| \\ &\leqslant \| \boldsymbol{A}_i \|^l \, \overline{\boldsymbol{w}}_i + \frac{1 - \| \boldsymbol{A}_i \|^l}{1 - \| \boldsymbol{A}_i \|} \overline{\boldsymbol{w}}_i \\ &= \frac{1 - \| \boldsymbol{A}_i \|^{l+1}}{1 - \| \boldsymbol{A}_i \|} \overline{\boldsymbol{w}}_i \end{aligned}$$

Consequently, $\boldsymbol{z} \in \boldsymbol{W}_i^{l+1}$. Since $\boldsymbol{X}_i^{l+1} = \boldsymbol{X}_i \sim \boldsymbol{W}_i^{l+1}$, it can be obtained that

$$\boldsymbol{y} + \boldsymbol{v} = \boldsymbol{z} + \boldsymbol{x} \in \boldsymbol{X}_i$$

Hence, $\boldsymbol{y} \in \boldsymbol{X}_i^l$. □

Next, the upper bound of disturbances is regulated in the following theorem, such that recursive feasibility and closed-loop stability can be guaranteed.

Theorem 2.3 *For subsystem* $\mathcal{A}_i$, *if the upper bound of the perturbation term satisfies*

$$\overline{w}_i \leqslant \min\left\{\sqrt{\frac{\Pi_i \eta_i \varepsilon_i}{\Xi_i}}, \frac{\sqrt{\frac{\varepsilon_i - \overline{\varepsilon}_i + \overline{\lambda}(\boldsymbol{P}_i)\overline{\boldsymbol{x}}_i^2}{\overline{\lambda}(\boldsymbol{P}_i)}} - \overline{\boldsymbol{x}}_i}{\|\boldsymbol{A}_i\|^{N-1}}\right\}$$

where $0<\eta_i<1, \Pi_i = \frac{\lambda(\overline{\boldsymbol{Q}}_i)}{\overline{\lambda}(\boldsymbol{P}_i)} - \frac{\boldsymbol{\mu}_i}{\underline{\lambda}(\boldsymbol{P}_i)}, \Xi_i = \overline{\lambda}(\boldsymbol{P}_i) + \frac{\|(\boldsymbol{A}_i + \boldsymbol{B}_i \boldsymbol{K}_i)^{\mathrm{T}} \boldsymbol{P}_i\|^2}{\boldsymbol{\mu}_i}$. *and there exists a constant* μ_i *satisfying*

$$\max\left\{0, \frac{[\lambda(\overline{\boldsymbol{Q}}_i) - \overline{\lambda}(\boldsymbol{P}_i)]\underline{\lambda}(\boldsymbol{P}_i)}{\overline{\lambda}(\boldsymbol{P}_i)}\right\} < \boldsymbol{\mu}_i < \frac{\lambda(\overline{\boldsymbol{Q}}_i)\underline{\lambda}(\boldsymbol{P}_i)}{\overline{\lambda}(\boldsymbol{P}_i)}$$

then Problem 2.2 is recursively feasible. Moreover, the state of subsystem $\mathcal{A}_i$ *will enters the disturbance invariant set* $\psi_i = \{\boldsymbol{x}_i \in \mathbb{R}^{n_i} : \|\boldsymbol{x}_i\|_{\boldsymbol{P}_i}^2 \leqslant \eta_i \varepsilon_i\}$ *in finite time.*

2.3.5 *Example*

To verify the effectiveness of the proposed Algorithm 2, we consider the consensus problem of an NCS composed of three agents that has been presented in [11]. Each agent $\mathcal{A}_i (i=1,2,3)$ is regarded as a subsystem, which is described as

$$\boldsymbol{x}_i(k+1) = \begin{bmatrix} 1 & 0.1 & 0 & 0 \\ 0 & 1 & 0 & 0 \\ 0 & 0 & 1 & 0.1 \\ 0 & 0 & 0 & 1 \end{bmatrix} \boldsymbol{x}_i(k) + \begin{bmatrix} 0.05 & 0 \\ 1 & 0 \\ 0 & 0.05 \\ 0 & 1 \end{bmatrix} \boldsymbol{u}_i(k) + \boldsymbol{w}_i(k) \tag{2.50}$$

where the state vector is defined as $\boldsymbol{x}_i = [x_{i1}, x_{i2}, x_{i3}, x_{i4}]^{\mathrm{T}}$, x_{i1} and x_{i3} are the horizontal position s_i^x and vertical position s_i^y, x_{i2} is the horizontal velocity v_i^x of agent i and x_{i4} is the vertical velocity v_i^y; the input vector is defined as $\boldsymbol{u}_i = [u_{i1}, u_{i2}]^{\mathrm{T}}$, u_{i1} and u_{i2} are the horizontal force u_i^x and vertical force u_i^y.

The constraints on control input and the perturbation term are given by $\|\boldsymbol{u}_i(k)\|_\infty \leqslant 0.15$ and $\|\boldsymbol{w}_i(k)\| \leqslant 0.009$. The parameters in Problem 2.2 are chosen as $N=15, \varepsilon_i = 2.4, \overline{\varepsilon}_i = 1.4, \sigma_1 = 0.9, \sigma_2 = 0.8, \sigma_3 = 0.8, \mathcal{N}_1 = \{2,3\}$, $\mathcal{N}_2 = \{1,3\}, \mathcal{N}_3 = \{1,2\}, \boldsymbol{Q}_i = \boldsymbol{I}_{4\times4}, \boldsymbol{R}_i = 5\boldsymbol{I}_{2\times2}, \boldsymbol{Q}_{ij} = [1,0,0,0;0,0,0,0;0,0,1,$

0;0,0,0,0]. The initial states satisfy $\boldsymbol{x}_1(0)=[1,0,-1,0]^{\mathrm{T}}$, $\boldsymbol{x}_2(0)=[1,0,1,0]^{\mathrm{T}}$, $\boldsymbol{x}_3(0)=[-1,0,-1,0]^{\mathrm{T}}$.

By solving the Riccati equation, the terminal weighting matrix and state feedback control law satisfying Lemma 2.2 are obtained as follows.

$$\boldsymbol{P}_i=\begin{bmatrix}12.5862 & 2.2913 & 0 & 0\\ 2.2913 & 3.2693 & 0 & 0\\ 0 & 0 & 12.5862 & 2.2913\\ 0 & 0 & 2.2913 & 3.2693\end{bmatrix}$$

$$\boldsymbol{K}_i=\begin{bmatrix}-0.3424 & -0.4309 & 0 & 0\\ 0 & 0 & -0.3424 & -0.4309\end{bmatrix}$$

The positions of three agents under Algorithm 2 and time-triggered dual-mode DMPC are depicted in Fig. 2.6. It is shown that the closed-loop behaviors under two control strategies are analogous. The horizontal velocities and vertical velocities of three subsystems are presented in Fig. 2.7. The horizontal forces and vertical forces are presented in Fig. 2.8. It can be observed that for both approaches, the implemented inputs satisfy the constraints. The NCS achieves consensus w. r. t. velocity and position variables. Figure 2.9 displays

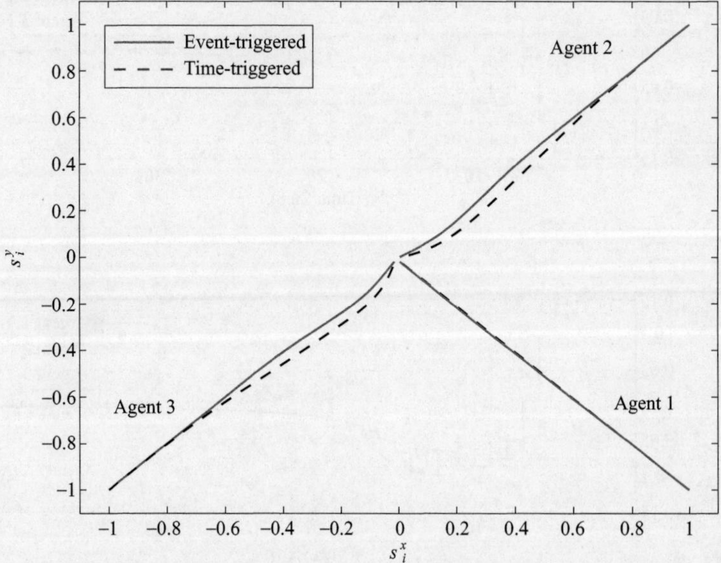

Fig. 2.6 Trajectories of three agents under event-triggered dual-mode DMPC and time-triggered dual-mode DMPC

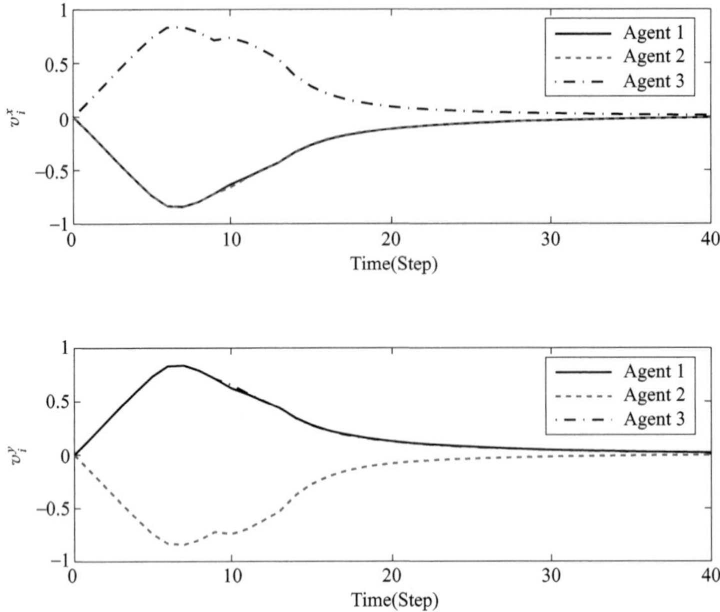

Fig. 2. 7 Horizontal velocities and vertical velocities of three agents

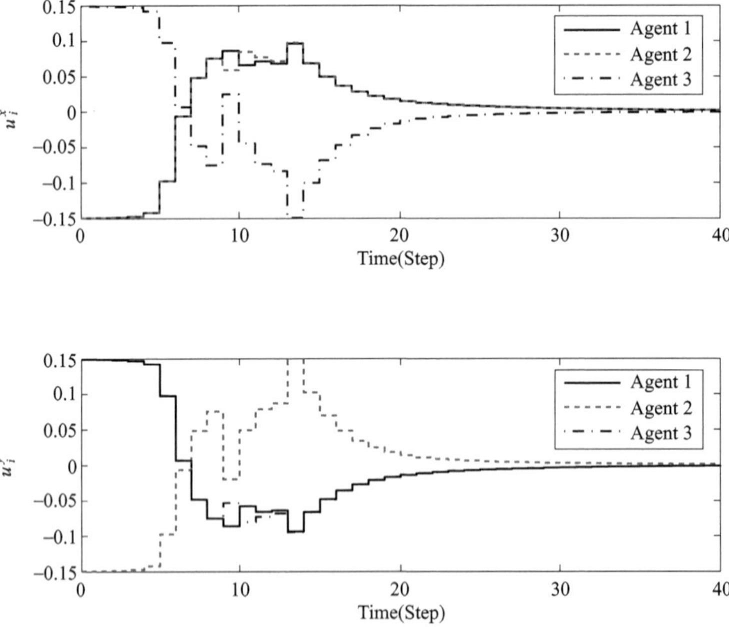

Fig. 2. 8 Horizontal forces and vertical forces of three agents

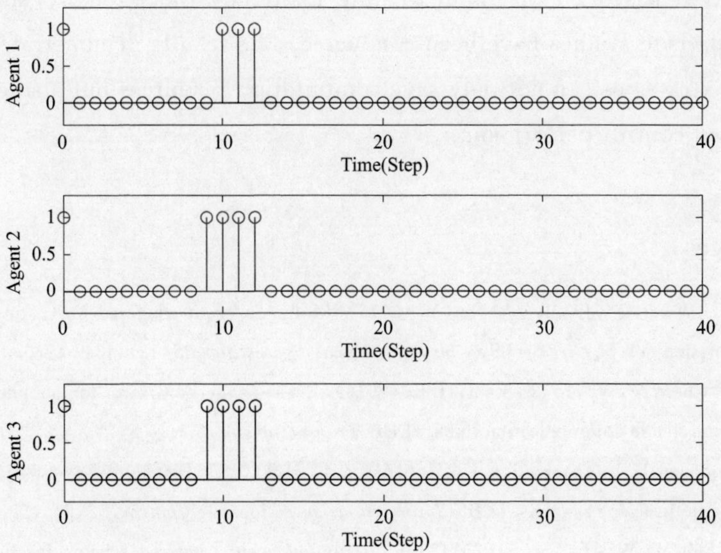

Fig. 2.9 Triggering instants for three agents

the triggering instants for three agents under proposed Algorithm 2, in which value 1 represents that Problem 2.2 is triggered and vice versa. It can be observed that Problem 2.2 is solved in an aperiodic manner. The state of the agent $\mathcal{A}_i$ stays in the invariant set ϕ_i in the last 20 steps. During this interval, the local state feedback controller is applied. The simulation results indicate that the proposed event-triggered DMPC schemes results in acceptable control performance, while saving the computation and communication consumption.

2.4 Discussion

In this chapter, the DMPC for NCSs with event-triggered computation has been studied. Section 2.2 proposes a cooperative event-triggering condition, which relies on the real-time information from both the local subsystem and the neighboring subsystems. Section 2.3 proposes a dual-mode DMPC strategy with a decentralized event-triggered computation mechanism. The triggering condition is parameterized at the local information only. Further improvement on resource reduction is achieved, resorting to the setting of terminal condition that copes with the system disturbances or the uncertainty arising in the aperiodic information exchanges. The closed-loop analysis for both strategies has been derived based on the input-to-state Lyapunov tools. Finally, simulations and comparison studies have been conducted. The results demonstrate that the designed strategies can not only save computation resources but also guarantee the desired control performance.

References

1. Zou, Y., Wang, Q., Jia, T., & Niu, Y. (2016). Multirate event-triggered MPC for NCSs with transmission delays. *Circuits, Systems, and Signal Processing*, *35*(12), 4249-4270.
2. Li, H., Chen, Z., Wu, L., & Lam, H. K. (2016). Event-triggered control for nonlinear systems under unreliable communication links. *IEEE Transactions on Fuzzy Systems*. *25*(4), 813-824.
3. Dimarogonas, D. V., Frazzoli, E., & Johansson, K. H. (2011). Distributed event-triggered control for multi-agent systems. *IEEE Transactions on Automatic Control*, *57*(5), 1291-1297.
4. Li, L., Ho, D. W., & Xu, S. (2014). A distributed event-triggered scheme for discrete-time multi-agent consensus with communication delays. *IET Control Theory and Applications*, 8(10), 830-837.
5. Zhu, W., & Jiang, Z. P. (2014). Event-based leader-following consensus of multi-agent systems with input time delay. *IEEE Transactions on Automatic Control*, *60*(5), 1362-1367.
6. Hashimoto, K., Adachi, S., & Dimarogonas, D. V. (2015). Distributed aperiodic model predictive control for multi-agent systems. *IET Control Theory and Applications*, *9*(1), 10-20.
7. Li, H., Yan, W., Shi, Y., & Wang, Y. (2015). Periodic event-triggering in distributed receding horizon control of nonlinear systems. *Systems and Control Letters*, *86*, 16-23.
8. Zou, Y., Su, X., & Niu, Y. (2016). Event-triggered distributed predictive control for the cooperation of multi-agent systems. *IET Control Theory and Applications*, *11*(1), 10-16.
9. Su, X., Zou, Y., Niu, Y., & Jia, T. (2016). Event-triggered distributed predictive control for constrained large-scale linear systems. In *2016 35th Chinese Control Conference (CCC)* (pp. 4253-4258). IEEE.
10. Jiang, Z. P., & Wang, Y. (2001). Input-to-state stability for discrete-time nonlinear systems. *Automatica*, *37*(6), 857-869.
11. Wang, C., & Ong, C. J. (2010). Distributed model predictive control of dynamically decoupled systems with coupled cost. *Automatica*, *46*(12), 2053-2058.
12. Chen, H., & Allgöwer, F. (1998). A quasi-infinite horizon nonlinear model predictive control scheme with guaranteed stability. *Automatica*, *34*(10), 1205-1217.

Chapter 3
DMPC of Networked Systems with Event-Triggered Communication

3.1 Introduction

The introduction of communication network, especially wireless network, brings benefits to distributed control but also leads to new challenges. Firstly, information exchange through the network is critical to realize a global cooperation objective through a distributed control, especially by the one in an iterative scheme [1,2]. However, the actual wireless nodes cannot afford the mass communication burden due to constrained battery power supply and network bandwidth, which may result in communication interruption and thus reducing control performance. Besides, information transmission consumes more resources than the processing, and data reception spends about the same order of magnitude of energy as the transmission [3]. Therefore, the design of the control approach to guarantee desired control performance with less communication usage has been an important research issue.

Event-triggered control strategies have been widely studied to reduce communication cost [4-7], where every controller is activated to receive its neighbors' states and broadcast its updated state to neighboring subsystems only when the predefined event occurs. For the design of DMPC with event-triggered communication, the main challenge lies in how to reduce the influences of state deviation caused by asynchronous information exchange on the system properties.

In Chap. 2, the DMPC problem of NCSs with event-triggered computation

Y. Zou and S. Li, *Distributed Cooperative Model Predictive Control of Networked Systems*,
https://doi.org/10.1007/978-981-19-6084-0_3

has been studied. However, the local controllers are required to exchange information with its neighboring controllers at each time step and it may results in unnecessary communication consumption. In this chapter, a DMPC algorithm with event-triggered communication for linear discrete-time NCSs is introduced. Firstly, in order to reduce communication load in a substantial extent, a triggering condition is constructed with considering to decrease the constructed Lyapunov function along the triggering instants rather than the sampling instants, that is, the controller is activated to solve the formulated FHOCP and communication with its neighbors only when the Lyapunov function is greater than the value at the most recent event-triggering instant. Secondly, the set level of the terminal constraint is time-varying between the maximal control invariant set and the target terminal set to expand the feasible region while remaining a satisfied control performance. The main results of this chapter have been published in [9].

The remainder of this chapter is organized as follows. Section 3.2 describes the control objective and states the formulated event-triggered DMPC problem. In Section 3.3, the event-triggering condition is derived and the DMPC algorithm based on this event-triggered mechanism is proposed. Section 3.4 further analyzes the recursive feasibility of the proposed algorithm and the closed-loop stability. Its effectiveness is verified in Section 3.5 by a numerical simulation. Finally, the conclusion is drawn in Section 3.6.

Notations: Throughout the chapter, $\mathbb{R}^n$ denotes the n -dimensional Euclidean space. The notation $\boldsymbol{P}\succ 0$ means that matrix $\boldsymbol{P}$ is a positive definite matrix. The symbol $*$ denotes the symmetric term in a symmetric matrix. The subscript i represents the ith subsystem. The superscripts T and -1 represent the matrix transposition and inverse, respectively. we use $\|\cdot\|$ to represent the Euclidean norm of a vector $\boldsymbol{x}\in\mathbb{R}^n$ and also the induced norm of a matrix.

3.2 Optimization Problem Formulation

Consider a NCS composed of M subsystems, where the dynamics of each subsystem is described as

$$\boldsymbol{x}_i(k+1)=\boldsymbol{A}_i\boldsymbol{x}_i(k)+\boldsymbol{B}_i\boldsymbol{u}_i(k) \tag{3.1}$$

where $i=1,\ldots,M$, $\boldsymbol{x}_i(k)\in\mathbb{R}^{n_i}$ and $\boldsymbol{u}_i(k)\in\mathbb{R}^{m_i}$ are the state and control input of subsystem i at time instant k, respectively. $\boldsymbol{A}_i\in\mathbb{R}^{n_i\times n_i}$ is the system matrix and $\boldsymbol{B}_i\in\mathbb{R}^{n_i\times m_i}$ is the input matrix. Besides, the state and control input are limited by $\boldsymbol{x}_i(k)\in\mathbb{X}_i$ and $\boldsymbol{u}_i(k)\in\mathbb{U}_i$, where both $\mathbb{X}_i$ and $\mathbb{U}_i$ are compact sets containing the origin.

This chapter mainly considers subsystems which are spatially distributed but coupled by the cooperation objective. As is shown in Fig. 3.1, an event-triggered MPC controller is applied to every subsystem, which solves a FHOCP and broadcasts system information to its neighbors only if an event occurs. Denote r_i as the rth triggering instant of subsystem i At time r_i, the following optimization problem is solved by the controller i based on its current state measurement $\boldsymbol{x}_i(r_i)$.

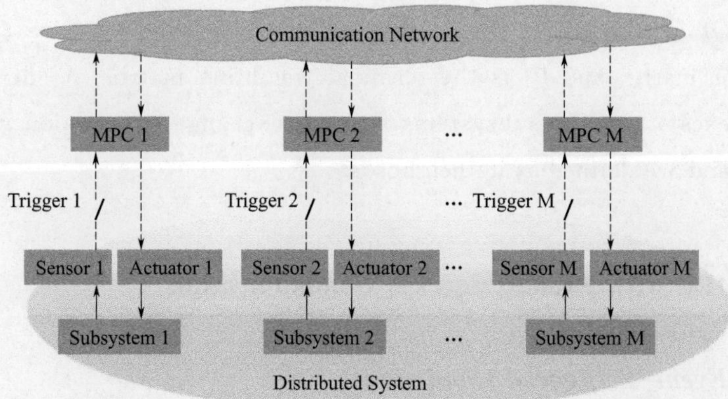

Fig. 3.1 Structure of DMPC with event-triggered communication: the solid lines indicate that the information is transmitted periodically, and the dotted lines indicate that the information transmission is driven by event triggers

Problem 3.1

$$\begin{aligned}
&\min_{\boldsymbol{u}_i(r_i)} J_i[\boldsymbol{x}_i(r_i),\boldsymbol{u}_i(r_i),\boldsymbol{x}_j(r_i)]\\
&=\min_{\boldsymbol{u}_i(r_i)}\sum_{l=0}^{N-1}\{L_i[\tilde{\boldsymbol{x}}_i(r_i+l\,|\,r_i),\tilde{\boldsymbol{u}}_i(r_i+l\,|\,r_i)]+\\
&\sum_{j\in\mathcal{N}_i}H_{ij}[\tilde{\boldsymbol{x}}_i(r_i+l\,|\,r_i),\tilde{\boldsymbol{x}}_j(r_i+l\,|\,r_i)]\}\\
&+V_{f_i}[\tilde{\boldsymbol{x}}_i(r_i+N\,|\,r_i)]
\end{aligned}$$

s. t.

$$\tilde{\boldsymbol{x}}_i(r_i+l+1|r_i)=\boldsymbol{A}_i\tilde{\boldsymbol{x}}_i(r_i+l|r_i)+\boldsymbol{B}_i\tilde{\boldsymbol{u}}_i(r_i+l|r_i), l=0,\ldots,N-1$$
$$\tilde{\boldsymbol{x}}_i(r_i+l|r_i)\in\mathbb{X}_i, l=0,\ldots,N$$
$$\tilde{\boldsymbol{u}}_i(r_i+l|r_i)\in\mathbb{U}_i, l=0,\ldots,N-1$$
$$\tilde{\boldsymbol{x}}_i(r_i+N|r_i)\in\boldsymbol{X}_f^i$$

where

$$L_i[\tilde{\boldsymbol{x}}_i(r_i+l|r_i),\tilde{\boldsymbol{u}}_i(r_i+l|r_i)]\triangleq\|\tilde{\boldsymbol{x}}_i(r_i+l|r_i)\|_{\boldsymbol{Q}_i}^2+\|\tilde{\boldsymbol{u}}_i(r_i+l|r_i)\|_{\boldsymbol{R}_i}^2$$
$$H_{ij}[\tilde{\boldsymbol{x}}_i(r_i+l|r_i),\tilde{\boldsymbol{x}}_j(r_i+l|r_i)]\triangleq\|\tilde{\boldsymbol{x}}_i(r_i+l|r_i)-\tilde{\boldsymbol{x}}_j(r_i+l|r_i)\|_{\boldsymbol{Q}_{ij}}^2$$
$$V_{f_i}[\tilde{\boldsymbol{x}}_i(r_i+N|r_i)]\triangleq\|\tilde{\boldsymbol{x}}_i(r_i+N|r_i)\|_{\boldsymbol{P}_i}^2$$

where N denotes the prediction horizon, $\tilde{\boldsymbol{u}}_i(r_i+l|r_i)$ is the input sequence to be optimized by controller i, $\tilde{\boldsymbol{x}}_i(r_i+l|r_i)$ is the corresponding predicted state trajectory based on $\tilde{\boldsymbol{x}}_i(r_i|r_i)=\boldsymbol{x}_i(r_i)$, $\boldsymbol{X}_f^i$ denotes the target terminal set of subsystem i, $\boldsymbol{Q}_i$ and $\boldsymbol{R}_i$ are the positive definite weighting matrices, $\boldsymbol{Q}_{ij}$ is the cooperation matrix. and $\boldsymbol{P}_i$ is the terminal weighting matrix. $\mathcal{N}_i$ denotes the set of indexes of specific subsystems that can exchange information with subsystem i and are defined as its neighbors.

3.3 DMPC with Event-Triggered Communication

3.3.1 Event-Triggered Condition

In this section, we design the triggering condition based on the analysis of Lyapunov stability. Before proceeding, we introduce the following standard conditions.

Assumption 3.1 The stage function $L_i(\boldsymbol{x}_i,\boldsymbol{u}_i)$ is Lipschitz continuous in $\mathbb{X}_i\times\mathbb{U}_i$. That is, $\forall\boldsymbol{x}_{1_i}\in\mathbb{X}_i$ and $\boldsymbol{x}_{2_i}\in\mathbb{X}_i$, $\exists L_{Q_i}>0$, s. t.

$$\|L_i(\boldsymbol{x}_{1_i},\boldsymbol{u}_{1_i})-L_i(\boldsymbol{x}_{2_i},\boldsymbol{u}_{2_i})\|\leqslant L_{Q_i}\|\boldsymbol{x}_{1_i}-\boldsymbol{x}_{2_i}\|$$

Assumption 3.2 The running cost $H_{ij}(\boldsymbol{x}_i,\boldsymbol{x}_j)$ is Lipschitz continuous in $\mathbb{X}_i\times\mathbb{X}_j$, with Lipschitz constants L_{qx}^i and $L_{qx^a}^i$, respectively.

Assumption 3.3 There exists an invariant set $\Phi_i\triangleq\{\boldsymbol{x}_i|\boldsymbol{x}_i^{\mathrm{T}}\boldsymbol{P}_i\boldsymbol{x}_i\leqslant\sigma^i, \sigma^i>0\}\subset\mathbb{X}_i$

and a terminal set $\boldsymbol{X}_f^i \triangleq \{\boldsymbol{x}_i \mid \boldsymbol{x}_i^{\mathrm{T}} \boldsymbol{P}_i \boldsymbol{x}_i \leqslant \sigma_f^i, 0 < \sigma_f^i \leqslant \sigma^i\}$, such that $0 \in \boldsymbol{X}_f^i \subset \Phi_i \subset \mathbb{X}_i$.

Assumption 3.4 If $\boldsymbol{x}_i(k) \in \Phi_i$, there exists a local stabilizing controller $\boldsymbol{K}_{f_i} \boldsymbol{x}_i(k) \in \boldsymbol{U}_i$, in the sense that $\tilde{\boldsymbol{x}}_i(k+1 \mid k) = \boldsymbol{A}_i \boldsymbol{x}_i(k) + \boldsymbol{B}_i \boldsymbol{K}_{f_i}, \boldsymbol{x}_i(k) \in \Phi_i$. In addition, $\forall \boldsymbol{x}_i(k) \in \boldsymbol{X}_f^i$, we have $\tilde{\boldsymbol{x}}_i(k+1 \mid k) \in \boldsymbol{X}_f^i$ and the associated terminal function follows

$$V_{f_i}[\tilde{\boldsymbol{x}}_i(k+1 \mid k)] - V_{f_i}[\boldsymbol{x}_i(k)] \leqslant -L_i[\boldsymbol{x}_i(k), \boldsymbol{u}_i(k)] - \sum_{j \in N_i} H_{ij}[\boldsymbol{x}_i(k), \boldsymbol{x}_j(k)]$$

Assumption 3.5 The terminal cost $V_{f_i}(\boldsymbol{x}_i)$ is Lipschitz continuous in $\boldsymbol{X}_f^i$ with constant L_{f_i}.

Assumption 3.6 Ideal wireless communication channel is considered here, i. e. the issues of data loss, delays, latency and jitter are neglected.

For system (3.1), denote the series $\{r_i\}(r_i \in \mathbb{N})$ as the time instants to solve the optimization. At the triggering time r_i, the obtained optimal control inputs and optimal states are defined as $\boldsymbol{U}_i^*(r_i) = [\boldsymbol{u}_i^*(r_i \mid r_i), \ldots, \boldsymbol{u}_i^*(r_i + N - 1 \mid r_i)]$, $\boldsymbol{X}_i^*(r_i) = [\boldsymbol{x}_i(r_i), \tilde{\boldsymbol{x}}_i^*(r_i + 1 \mid r_i), \ldots, \tilde{\boldsymbol{x}}_i^*(r_i + N \mid r_i)]$. At sampling time $r_i + m\ (m = 1, \ldots, N-1)$, a control sequence $\overline{\boldsymbol{U}}_i(r_i + m)$ can be constructed based on $\boldsymbol{U}_i^*(r_i)$, see(3.2). Specifically, the corresponding inputs in $\boldsymbol{U}_i^*(r_i)$ are set as the predicted inputs during time interval $(r_i + m, r_i + N - 1)$. After time $r_i + N$, the feedback control law is employed to construct control inputs based on the predicted states. If the triggering condition is satisfied at time $r_i + m$, then Problem 3.1 is solved and the control inputs are updated; otherwise, employ the current control action.

$$\overline{\boldsymbol{u}}_i(r_i + t \mid r_i + m) = \begin{cases} \boldsymbol{u}_i^*(r_i + t \mid r_i), & t = 1, \ldots, N-1 \\ \boldsymbol{K}_{fi} \tilde{\boldsymbol{x}}_i(r_i + t \mid r_i + m), & t = N, \ldots, N+m-1 \end{cases} \tag{3.2}$$

If the initial state is feasible, the feasibility of the control inputs $\widetilde{\boldsymbol{U}}_i(r_i + m)$ can be proved by recursive procedure. With the above preliminaries, we give the following lemma.

Lemma 3.1 *Consider the system (3.1) with the optimization problem as in Problem 3.1. Suppose that Assumptions 3.1-3.6 hold, Problem 3.1 of the ith subsystem is computed at time r_i, and the minimal index is $J_i^*(r_i)$. Let*

$\overline{J}_i(r_i+t)$ *denote the performance index under the feasible control inputs as in (3.2). If subsystem i receives no information from its neighboring subsystems, the error between* $\overline{J}_i(r_i+t)$ *and* $J_i^*(r_i)$ *is bounded by*

$$\overline{J}_i(r_i+t)-J_i^*(r_i)\leqslant L_{A1}(t)\|\boldsymbol{x}_i(r_i+t)-\tilde{\boldsymbol{x}}_i^*(r_i+t|r_i)\|-\sigma_{a_i}(t) \tag{3.3}$$

Otherwise,

$$\begin{aligned}\overline{J}_i(r_i+t)-J_i^*(r_i)\leqslant &L_{A2}(t)\|\boldsymbol{x}_i(r_i+t)-\tilde{\boldsymbol{x}}_i^*(r_i+t|r_i)\|-\sigma_{a_i}(t)\\&+\sum_{l=t}^{N-1}\sum_{j\in\mathcal{N}_i}L_{qx^a}^i\|\tilde{\boldsymbol{x}}_j(r_i+l|r_j^u)-\tilde{\boldsymbol{x}}_j(r_i+l|r_j)\|\end{aligned} \tag{3.4}$$

where

$$L_{\boldsymbol{A}1}(t)\triangleq L_{Q_i}\sum_{l=t}^{N-1}\|\boldsymbol{A}_i\|^{l-t}+L_{f_i}\|\boldsymbol{A}_i\|^{N-t}+|\mathcal{N}_i|L_{qx}^i\sum_{l=t}^{N-1}\|\boldsymbol{A}_i\|^{l-t}$$

$$\begin{aligned}\sigma_{a_i}(t)\triangleq &\sum_{l=0}^{t-1}\sum_{j\in\mathcal{N}_i}H_{ij}[\tilde{\boldsymbol{x}}_i^*(r_i+l|r_i),\tilde{\boldsymbol{x}}_j(r_i+l|r_j)]+\\&\sum_{l=0}^{t-1}L_i[\tilde{\boldsymbol{x}}_i^*(r_i+l|r_i),\boldsymbol{u}_i^*(r_i+l|r_i)]\end{aligned}$$

$$L_{\boldsymbol{A}2}(t)\triangleq L_{Q_i}\sum_{l=t}^{N-1}\|\boldsymbol{A}_i\|^{l-t}+L_{f_i}\|\boldsymbol{A}_i\|^{N-t}$$

where $r_j=\arg\min_{r_j}\{r_i-r_j>0\}$ *is the most recent triggering time of the jth subsystem at time* r_i, *and* $r_j^u=\arg\min_{r_j}\{r_i+t-r_j>0\}$ *is that of the jth subsystem at current time* r_i+t. $|\mathcal{N}_i|$ *indicates the number of its neighbors.*

Proof In this proof, we will prove that for any $t=1,\ldots,N-1$, the inequalities (3.3) or (3.4) always holds. First the inequality (3.3) is proved, and then the inequality (3.4) is proved by the same procedure.

First, suppose subsystem i receives no information from its neighboring subsystems between time r_i and r_i+t. Therefore, the prediction of the future states of its neighboring subsystems are obtained based on the states at the same time. We then have that the error between the cost functions at the current time instants and the most recent triggering time instants is:

$$\overline{J}_i(r_i+t)-J_i^*(r_i)$$

$$
\begin{aligned}
&=\sum_{l=0}^{N-1}\{L_i[\tilde{\boldsymbol{x}}_i(r_i+t+l\,|\,r_i+t),\bar{\boldsymbol{u}}_i(r_i+t+l\,|\,r_i+t)]\\
&\quad+\sum_{j\in\mathcal{N}_i}H_{ij}[\tilde{\boldsymbol{x}}_i(r_i+t+l\,|\,r_i+t),\tilde{\boldsymbol{x}}_j(r_i+t+l\,|\,r_j)]\}+V_{f_i}[\tilde{\boldsymbol{x}}_i(r_i+N+t\,|\,r_i+t)]\\
&\quad-\sum_{l=0}^{N-1}\Big\{L_i[\tilde{\boldsymbol{x}}_i^*(r_i+l\,|\,r_i),\boldsymbol{u}_i^*(r_i+l\,|\,r_i)]+\sum_{j\in\mathcal{N}_i}H_{ij}[\tilde{\boldsymbol{x}}_i^*(r_i+l\,|\,r_i),\\
&\quad\tilde{\boldsymbol{x}}_j(r_i+l\,|\,r_j)]\Big\}\\
&\quad-V_{f_i}[\tilde{\boldsymbol{x}}_i^*(r_i+N\,|\,r_i)]\\
&=\sum_{l=t}^{N-1}\{L_i[\tilde{\boldsymbol{x}}_i(r_i+l\,|\,r_i+t),\bar{\boldsymbol{u}}_i(r_i+l\,|\,r_i+t)]-L_i[\tilde{\boldsymbol{x}}_i^*(r_i+l\,|\,r_i),\\
&\quad\boldsymbol{u}_i^*(r_i+l\,|\,r_i)]\}\\
&\quad+\sum_{l=N}^{N+t-1}\{L_i[\tilde{\boldsymbol{x}}_i(r_i+l\,|\,r_i+t),\tilde{\boldsymbol{u}}_i(r_i+l\,|\,r_i+t)]\\
&\quad+\sum_{j\in\mathcal{N}_i}H_{ij}[\tilde{\boldsymbol{x}}_i(r_i+l\,|\,r_i+t),\tilde{\boldsymbol{x}}_j(r_i+l\,|\,r_j)]\}\\
&\quad+V_{f_i}[\tilde{\boldsymbol{x}}_i(r_i+N+t\,|\,r_i+t)]-V_{f_i}[\tilde{\boldsymbol{x}}_i^*(r_i+N\,|\,r_i)] \qquad (3.5)\\
&\quad+\sum_{l=t}^{N-1}\sum_{j\in\mathcal{N}_i}H_{ij}[\tilde{\boldsymbol{x}}_i(r_i+l\,|\,r_i+t),\tilde{\boldsymbol{x}}_j(r_i+l\,|\,r_j)]\\
&\quad-\sum_{l=0}^{N-1}\sum_{j\in\mathcal{N}_i}H_{ij}[\tilde{\boldsymbol{x}}_i^*(r_i+l\,|\,r_i),\tilde{\boldsymbol{x}}_j(r_i+l\,|\,r_j)]\\
&\quad-\sum_{l=0}^{t-1}L_i[\tilde{\boldsymbol{x}}_i^*(r_i+l\,|\,r_i),\tilde{\boldsymbol{u}}_i^*(r_i+l\,|\,r_i)]
\end{aligned}
$$

Based on (3.2), we have $\bar{\boldsymbol{u}}_i(r_i+l\,|\,r_i+t)=\boldsymbol{u}_i^*(r_i+l\,|\,r_i)$ for $l=t,\ldots,N-1$. According to Assumption 3.1, it can be obtained that the error for the stage cost is bounded by

$$
\begin{aligned}
&\sum_{l=t}^{N-1}\{L_i[\tilde{\boldsymbol{x}}_i(r_i+l\,|\,r_i+t),\tilde{\boldsymbol{u}}_i(r_i+l\,|\,r_i+t)]-L_i[\tilde{\boldsymbol{x}}_i^*(r_i+l\,|\,r_i),\\
&\tilde{\boldsymbol{u}}_i^*(r_i+l\,|\,r_i)]\}\\
&\leqslant L_{Q_i}\sum_{l=t}^{N-1}\|\tilde{\boldsymbol{x}}_i(r_i+l\,|\,r_i+t)-\tilde{\boldsymbol{x}}_i^*(r_i+l\,|\,r_i)\| \qquad (3.6)\\
&=L_{Q_i}\sum_{l=t}^{N-1}\|\boldsymbol{A}_i^{l-t}[\boldsymbol{x}_i(r_i+t)-\tilde{\boldsymbol{x}}_i^*(r_i+t\,|\,r_i)]\|\\
&\leqslant L_{Q_i}\sum_{l=t}^{N-1}\|\boldsymbol{A}_i\|^{l-t}\|\boldsymbol{x}_i(r_i+t)-\tilde{\boldsymbol{x}}_i^*(r_i+t\,|\,r_i)\|
\end{aligned}
$$

Based on Assumption 3.4, we have

$$\begin{aligned}&L_i[\tilde{\boldsymbol{x}}_i(r_i+N|r_i+t),\bar{\boldsymbol{u}}_i(r_i+N|r_i+t)]\\&+\sum_{j\in\mathcal{N}_i}H_{ij}[\tilde{\boldsymbol{x}}_i(r_i+N|r_i+t),\tilde{\boldsymbol{x}}_j(r_i+N|r_j)]\\&+V_{f_i}[\tilde{\boldsymbol{x}}_i(r_i+1+N|r_i+t)]\\&\leqslant V_{f_i}[\tilde{\boldsymbol{x}}_i(r_i+N|r_i+t)]\end{aligned}\tag{3.7}$$

Similarly, we can obtain the following inequations

$$\begin{aligned}&L_i[\tilde{\boldsymbol{x}}_i(r_i+N+h|r_i+t),\bar{\boldsymbol{u}}_i(r_i+N+h|r_i+t)]\\&+\sum_{j\in N_i}H_{ij}[\tilde{\boldsymbol{x}}_i(r_i+N+h|r_i+t),\tilde{\boldsymbol{x}}_j(r_i+N+h|r_j)]\\&+V_{f_i}[\tilde{\boldsymbol{x}}_i(r_i+h+N+1|r_i+t)]\\&\leqslant V_{f_i}[\tilde{\boldsymbol{x}}_i(r_i+N+h|r_i+t)]\end{aligned}\tag{3.8}$$

with $h=0,\ldots,t-1$.

Summing all the inequalities in (3.8), we have

$$\begin{aligned}&\sum_{l=N}^{N+t-1}\{L_i[\tilde{\boldsymbol{x}}_i(r_i+l|r_i+t),\tilde{\boldsymbol{u}}_i(r_i+l|r_i+t)]\\&+\sum_{j\in\mathcal{N}_i}H_{ij}[\tilde{\boldsymbol{x}}_i(r_i+l|r_i+t),\tilde{\boldsymbol{x}}_j(r_i+l|r_j)]\}\\&+V_{f_i}[\tilde{\boldsymbol{x}}_i(r_i+N+t|r_i+t)]\\&\leqslant V_{f_i}[\tilde{\boldsymbol{x}}_i(r_i+N|r_i+t)]\end{aligned}$$

Based on the discussion above and Assumption 3.5, we can obtain the following inequality as:

$$\begin{aligned}&\sum_{l=N}^{N+t-1}\{L_i[\tilde{\boldsymbol{x}}_i(r_i+l|r_i+t),\tilde{\boldsymbol{u}}_i(r_i+l|r_i+t)]\\&+\sum_{j\in\mathcal{N}_i}H_{ij}[\tilde{\boldsymbol{x}}_i(r_i+l|r_i+t),\tilde{\boldsymbol{x}}_j(r_i+l|r_j)]\}\\&+V_{f_i}[\tilde{\boldsymbol{x}}_i(r_i+N+t|r_i+t)]-V_{f_i}[\tilde{\boldsymbol{x}}_i^*(r_i+N|r_i)]\\&\leqslant V_{f_i}[\tilde{\boldsymbol{x}}_i(r_i+N|r_i+t)]-V_{f_i}[\tilde{\boldsymbol{x}}_i^*(r_i+N|r_i)]\\&\leqslant L_{f_i}\|\tilde{\boldsymbol{x}}_i(r_i+N|r_i+t)-\tilde{\boldsymbol{x}}_i^*(r_i+N|r_i)\|\end{aligned}\tag{3.9}$$

$$\leqslant L_{f_i}\|\boldsymbol{A}_i\|^{N-t}\|\boldsymbol{x}_i(r_i+t)-\tilde{\boldsymbol{x}}_i^*(r_i+t|r_i)\|$$

Since there is no information interaction between the neighbors during time interval (r_i, r_i+t), the prediction states of the neighboring subsystems are the same during time interval (r_i+t, r_i+N-1). Then according to the Assumption 3.2, we have that, for $l=t,\ldots,N-1$

$$H_{ij}[\tilde{\boldsymbol{x}}_i(r_i+l|r_i+t),\tilde{\boldsymbol{x}}_j(r_i+l|r_j)]-H_{ij}[\tilde{\boldsymbol{x}}_i^*(r_i+l|r_i),\tilde{\boldsymbol{x}}_j(r_i+l|r_j)]$$
$$\leqslant L_{qx}^i\|\tilde{\boldsymbol{x}}_i(r_i+l|r_i+t)-\tilde{\boldsymbol{x}}_i^*(r_i+l|r_i)\|$$

Therefore,

$$\begin{aligned}
&\sum_{l=t}^{N-1}\sum_{j\in\mathcal{N}_i}H_{ij}[\tilde{\boldsymbol{x}}_i(r_i+l|r_i+t),\tilde{\boldsymbol{x}}_j(r_i+l|r_j)]\\
&-\sum_{l=0}^{N-1}\sum_{j\in\mathcal{N}_i}H_{ij}[\tilde{\boldsymbol{x}}_i^*(r_i+l|r_i),\tilde{\boldsymbol{x}}_j(r_i+l|r_j)] \qquad (3.10)\\
&\leqslant L_{qx}^i\sum_{l=t}^{N-1}\sum_{j\in\mathcal{N}_i}\|\tilde{\boldsymbol{x}}_i(r_i+l|r_i+t)-\tilde{\boldsymbol{x}}_i^*(r_i+l|r_i)\|\\
&-\sum_{l=0}^{t-1}\sum_{j\in\mathcal{N}_i}H_{ij}[\tilde{\boldsymbol{x}}_i^*(r_i+l|r_i),\tilde{\boldsymbol{x}}_j(r_i+l|r_j)]\\
&\leqslant|\mathcal{N}_i|\cdot L_{qx}^i\sum_{l=t}^{N-1}\|\boldsymbol{A}_i\|^{l-t}\|\boldsymbol{x}_i(r_i+t)-\tilde{\boldsymbol{x}}_i^*(r_i+t|r_i)\|\\
&-\sum_{l=0}^{t-1}\sum_{j\in\mathcal{N}_i}H_{ij}[\tilde{\boldsymbol{x}}_i^*(r_i+l|r_i),\tilde{\boldsymbol{x}}_j(r_i+l|r_j)]
\end{aligned}$$

Summarizing the inequality (3.6), (3.9) and (3.10), we have that the performance error is bounded and the inequality (3.3) is proved.

If the jth subsystem sends its current information to the ith subsystem at r_j^u, we can obtain the following inequality similarly:

$$\begin{aligned}
&\sum_{l=t}^{N-1}\sum_{j\in\mathcal{N}_i}H_{ij}[\tilde{\boldsymbol{x}}_i(r_i+l|r_i+t),\tilde{\boldsymbol{x}}_j(r_i+l|r_j^u)]\\
&-\sum_{l=0}^{N-1}\sum_{j\in\mathcal{N}_i}H_{ij}[\tilde{\boldsymbol{x}}_i^*(r_i+l|r_i),\tilde{\boldsymbol{x}}_j(r_i+l|r_j)] \qquad (3.11)\\
&\leqslant\sum_{l=t}^{N-1}\sum_{j\in\mathcal{N}_i}L_{qx^a}^i\|\tilde{\boldsymbol{x}}_j(r_i+l|r_j^u)-\tilde{\boldsymbol{x}}_j(r_i+l|r_j)\|\\
&-\sum_{l=0}^{t-1}\sum_{j\in\mathcal{N}_i}H_{ij}[\tilde{\boldsymbol{x}}_i^*(r_i+l|r_i),\tilde{\boldsymbol{x}}_j(r_i+l|r_j)]
\end{aligned}$$

Summarizing the inequality (3.6),(3.9) and (3.11),the inequality (3.4) is proved. □

From Lemma 3.1,we have that the performance difference involving the optimal states calculated at themost recent triggering time and the current states. If we execute the following conditions

$$\|\boldsymbol{x}_i(r_i+t)-\tilde{\boldsymbol{x}}_i^*(r_i+t|r_i)\|\leqslant\beta\sigma_{a_i}(t)/L_{\boldsymbol{A}1}(t)$$

and

$$\sum_{l=t}^{N-1}\sum_{j\in\mathcal{N}_i}(L_{qx^a}^i\|\tilde{\boldsymbol{x}}_j(r_i+l|r_j^u)-\tilde{\boldsymbol{x}}_j(r_i+l|r_j)\|)$$
$$+L_{\boldsymbol{A}2}(t)\|\boldsymbol{x}_i(r_i+t)-\tilde{\boldsymbol{x}}_i^*(r_i+t|r_i)\|\leqslant\beta\sigma_{a_i}(t)$$

with $0<\beta<1$, we get $\overline{J}_i(r_i+t)<J_i^*(r_i)$. Thus,the event triggering condition of subsystem i is

$$\|\boldsymbol{x}_i(r_i+t)-\tilde{\boldsymbol{x}}_i^*(r_i+t|r_i)\|\geqslant\beta\sigma_{a_i}(t)/L_{\boldsymbol{A}1}(t)\tag{3.12}$$

or

$$\sum_{l=t}^{N-1}\sum_{j\in\mathcal{N}_i}(L_{qx^a}^i\|\tilde{\boldsymbol{x}}_j(r_i+l|r_j^u)-\tilde{\boldsymbol{x}}_j(r_i+l|r_j)\|)$$
$$+L_{\boldsymbol{A}2}(t)\|\boldsymbol{x}_i(r_i+t)-\tilde{\boldsymbol{x}}_i^*(r_i+t|r_i)\|\geqslant\beta\sigma_{a_i}(t)\tag{3.13}$$

The triggering conditions (3.12)and(3.13) determine whether the sensor sends the current information to the controller,and verify if there is information interaction between subsystems. Here is the overall update strategy. At the initial time,the system states are transmitted from the sensor to the controller,and information exchange is executed between neighboring subsystems such that the initial control inputs and initial prediction of states can be obtained. Then the controller transforms the predicted state trajectory to the triggering checking device. At the next time instants,the current states are sampled and the conditions (3.12)and (3.13)are checked. Either(3.12)or(3.13)is satisfied,the sensor sends states to the controller and Problem 3.1 is solved,then the obtained information is transformed to its neighboring subsystems. If not, the sensor would not sample,the subsystem would not be updated and the control action calculated at the most recent time instant is applied to the system. The above procedure is repeated at each time step.

Remark 3.1 Note that the threshold changes over time, and may raise from the triggered time instants since $\sigma_{a_i}(t)$ is increasing but $L_{A1}(t)$ and the item related to the states of its neighbors on the left side of inequality (3.13) decreases over time. However, the threshold shows a decreasing trend as the states of subsystems converge to the steady-state. Since the threshold indicates the tolerance of the maximum difference in the state prediction without violating the stability conditions, the triggering condition (3.12) or (3.13) is precisely the way we desired. In addition, the subsystem is triggered only when the performance value exceeds the value at the last triggering time instant, which results in less triggering frequency compared with the strategy in [12] where the transfer occurs if the cost function is not continuously decreasing at each sampling time instant. However, since the monotonicity is relaxed in the designed strategy, it may take longer to reach the steady-state.

Remark 3.2 In the DMPC implementation, we assume that the adjacent predicted states satisfy $\tilde{\boldsymbol{x}}_j(r_i+N+l\,|\,r_j)=\tilde{\boldsymbol{x}}_j(r_i+N\,|\,r_j), l=1,\ldots,N$.

3.3.2 DMPC Algorithm with Event-Triggered Communication

This section presents an improved scheme to enlarge the feasible area of the optimization, and then introduces the overall control algorithm.

Since the prediction difference increase over time, it is not appropriate for an long prediction horizon. However, a short prediction horizon may result in the infeasibility of the control. What's more, there may be no feasible solution for the event-triggered DMPC proposed above due to the constraints or disturbances. Based on the above considerations, we refer to the strategy in [10] to expand the feasible area by optimizing the terminal constrained set when Problem 3.1 is infeasible.

Specifically, a target terminal set $\boldsymbol{X}_f^i$ is constructed first and the maximal control invariant ellipsoid set $\boldsymbol{\Phi}_i$ ($\boldsymbol{X}_f^i \subset \boldsymbol{\Phi}_i$) is calculated, where $\boldsymbol{X}_f^i$ and $\boldsymbol{\Phi}_i$ satisfy Assumptions 3.3 and 3.5. Solve the optimization problem as in Problem 3.2 to control the terminal set-valued state prediction $\tilde{\boldsymbol{x}}_i(r_i+N\,|\,r_i)$ into $\boldsymbol{X}_f^i$. In this way, the state can gradually converge to the terminal set as long as the initial state is feasible to $\boldsymbol{\Phi}_i$ [10]. Therefore, the feasibility of the optimization is guaranteed and the feasible area of the optimization is extended.

Problem 3.2

$$\min_{\alpha_i, \mathbf{U}_i^*(k)} J_i[\boldsymbol{x}_i(k), \boldsymbol{u}_i(k), \boldsymbol{x}_j(k)]$$

s.t.

$$0 \leqslant \alpha_i(k) \leqslant \alpha_i(k-1) \leqslant 1,$$

$$\tilde{\boldsymbol{x}}_i(k+l|k) = \boldsymbol{A}_i \tilde{\boldsymbol{x}}_i(k+l-1|k) + \boldsymbol{B}_i \tilde{\boldsymbol{u}}_i(k+l-1|k),$$

$$\tilde{\boldsymbol{x}}_i(k+l|k) \in \boldsymbol{X}_i, l=1,\ldots,N,$$

$$\tilde{\boldsymbol{u}}_i(k+l|k) \in \boldsymbol{U}_i, l=0,\ldots,N-1,$$

$$\tilde{\boldsymbol{x}}_i(k+N|k) \in (1-\alpha_i)\boldsymbol{X}_f^i + \alpha_i \boldsymbol{\Phi}_i,$$

The triggering condition is similar to (3.12)-(3.13). And the overall control strategy for each subsystem can be formulated as in Algorithm 1, taking the ith subsystem for example.

Algorithm 1 DMPC Algorithm with Event-triggered Communication

---

For subsystem i:

1: At $r_i=0, k=0$, Initialization. Set $\boldsymbol{X}_f^i$, compute $\boldsymbol{\Phi}_i$, measure $\boldsymbol{x}_i(0)$, and communicate with its neighboring subsystems, then solve Problem 3.1. Finally, exchange $\boldsymbol{X}_i^*(0)$. Let $k=k+1$.

2: At time k, sample $\boldsymbol{x}_i(k)$. If (3.12) or (3.13) is satisfied, turn to Step 4; else, turn to Step 3.

3: If $k=r_i+N$, turn to Step 4. Otherwise, apply the corresponding control input $\boldsymbol{u}_i^*(k|r_i)$ to the system, turn to Step 5.

4: Solve Problem 3.1 or Problem 3.2, apply the current control input $\boldsymbol{u}_i^*(k)$ to the local subsystem. send $\boldsymbol{X}_i^*(k)$ and turn to Step 6.

5: Wait for the next sampling, let $k=k+1$ and turn to Step 2.

6: Let $r_i=k, k=k+1$, and turn to Step 2.

Remark 3.3 If the subsystems are triggered synchronously, an iterative algorithm is employed to achieve the cooperation between agents.

Remark 3.4 Problem 3.2 is solved only if Problem 3.1 is infeasible. If Problem 3.1 is replaced by Problem 3.2 with the same parameters in the algorithm implementation, the triggering frequency is less than that of using Problem 3.1, but the values of the cost are corresponding larger.

3.4 Feasibility and Stability Analysis

This section analyses the feasibility and stability of the proposed DMPC control strategy, and proves that the closed-loop system is Lyapunov stable.

Theorem 3.1 *Consider the system (3.1), and assume that Assumptions 3.1-3.6 hold, the control inputs are obtained by Algorithm 1 under the triggering condition (3.12) and (3.13). If the initial state is feasible, then the state of the system can converge to the desired terminal set within a finite time step.*

Proof In the proposed event-triggered mechanism, if the predicted states at the prediction horizon $\tilde{\boldsymbol{x}}_i(k+N|k)$ exceeds the terminal set $\boldsymbol{X}_f^i$, Problem 3.2 is solved. Then according to (3.2), a feasible control input sequence can be constructed based on the optimal solution obtained at k, and the optimal terminal set can get closer to the target set. Hence, the control inputs are always recursively feasible. At the triggering time instant r_i, we can obtain the optimal cost function as $J_i^*(r_i)$. At the next time step r_i+1, with the triggering condition(3.12)-(3.13), we have that the performance value corresponding to the feasible input follows $\overline{J}_i(r_i+1) \leqslant J_i^*(r_i)$, and then we have $J_i^*(r_i+l) \leqslant J_i^*(r_i)$. That is, the system can be controlled to the terminal set gradually in finite time. Prove complete. □

Selection of the Parameters In the proposed algorithm, the system performance is affected by many parameters. Here we provide some guidelines for their selection.

- Parameters in the optimization problems: The prediction horizon length N can realize the trade-off between the system stability and quickness, and is also set as the maximum triggering time interval in the mentioned algorithm. Although the algorithm can guarantee the system performance with a shorter prediction horizon compared with the traditional MPC, the resource usage can not be saved if a large horizon N is chosen. Thus, the value of N should not be too large. Note that in this chapter, we assume that the control horizon and the prediction horizon are equal. In addition, the weighting matrices $\boldsymbol{Q}_i$, $\boldsymbol{Q}_{ij}$, $\boldsymbol{R}_i$ are symmetric positive definite, and the elements in them can be determined based on the actual demands. What's more, the matrix $\boldsymbol{P}_i$ and the feedback control gain $\boldsymbol{K}_{f_i}$ should be chosen to satisfy the Assumptions 3.3 and 3.4.
- Parameters in Assumptions 3.1-3.5: σ_i indicates the initial feasible set. σ_f^i in

the terminal set stands for the final convergence of the system, and we have that, the smaller the σ_f^i, the closer the system state is to the target point. Specifically, once the target terminal set is determined, σ_i can be obtained by employing the i-step controllable set [12]. Then, L_{Q_i}, L_{qx}^i, $L_{qx^a}^i$, L_{f_i}, $\boldsymbol{K}_{f_i}$ are calculated by trial-and-error method. The parameter β_i in the triggering condition follows $\beta_i \in (0,1)$, which plays an important part in the trade-off between system stability and communication frequency: the conditions (3.12) and (3.13) with a smaller β_i result in higher communication rates and a better system performance.

3.5 Example

In this section, the effectiveness of the Algorithm 1 is verified by simulations. A discrete LTI system composed of 3 agents with a cooperative objection is considered, where the parameters of the agents' dynamics follows:

$$\boldsymbol{A}_1=\begin{bmatrix}1 & 0.2\\ -0.5 & -0.4\end{bmatrix}, \boldsymbol{A}_2=\begin{bmatrix}1 & 0.2\\ -1.2 & -0.2\end{bmatrix}$$

$$\boldsymbol{A}_3=\boldsymbol{A}_2, \boldsymbol{B}_3=\boldsymbol{B}_2=\boldsymbol{B}_1=[0,0.2]^{\mathrm{T}}$$

The states and control inputs are constrained by $\|\boldsymbol{x}_i\|_2 \leqslant 5$, $\|\boldsymbol{u}_i\|_2 \leqslant 1$. Based on the mentioned parameter selection principles, the tuning parameters of each agent are set as $N=5$, $\boldsymbol{Q}_1=\boldsymbol{Q}_2=\boldsymbol{Q}_3=\boldsymbol{I}_2$, $\boldsymbol{R}_1=\boldsymbol{R}_2=\boldsymbol{R}_3=\boldsymbol{I}_1$, $\boldsymbol{Q}_{12}=\boldsymbol{Q}_{13}=0.5\boldsymbol{I}_2$, $\boldsymbol{Q}_{21}=\boldsymbol{Q}_{31}=\boldsymbol{I}_2$. To satisfy Assumptions 3.3 and 3.4, we set $\sigma_f^1=1$, $\sigma_f^2=1$, $\sigma_f^3=1$, and $\sigma^1=36$, $\sigma^2=33$, $\sigma^3=33$, and the matrices $\boldsymbol{P}_i$, $\boldsymbol{K}_i$ $(i=1,2,3)$ are obtained as:

$$\boldsymbol{P}_1=\begin{bmatrix}12.1548 & 2.0205\\ 2.0205 & 1.8282\end{bmatrix}, \quad \boldsymbol{P}_2=\boldsymbol{P}_3=\begin{bmatrix}9.7642 & 2.0592\\ 2.0592 & 1.769\end{bmatrix},$$

$$\boldsymbol{K}_1=[-0.2861, 0.0185], \quad \boldsymbol{K}_2=\boldsymbol{K}_3=[0.0643, -0.0011].$$

The other parameters are determined in a numerical way, such as, $L_{Q_1}=67$, $L_{Q_2}=47$, $L_{Q_3}=47$, $L_{f_1}=13$, $L_{f_2}=16$, $L_{f_3}=9$, $L_{qx}^1=19$, $L_{qx}^2=32$, $L_{qx}^3=3$, $L_{qx^a}^1=24$, $L_{qx^a}^2=31$, $L_{qx^a}^3=31$, $\beta_1=0.87$, $\beta_2=0.9$, $\beta_3=0.9$.

First, we execute the proposed Algorithm 1 at the feasible initial states $\boldsymbol{x}_1(0)=$

$[0.5, 0.5]^{\mathrm{T}}$, $\boldsymbol{x}_2(0) = [-1, -2]^{\mathrm{T}}$, $\boldsymbol{x}_3(0) = [0, 3]^{\mathrm{T}}$ with a small disturbances w_i, $\|w_i\| \leqslant 0.05$, $i = 1, 2, 3$. Then as shown in Figs. 3.2, 3.3, 3.4, 3.5 and 3.6, the states trajectories, the values of the cost function, the control inputs, and the triggering time instants obtained based on the designed event-triggered DMPC algorithm are compared with those of the traditional time-triggered DMPC as in [1] and the event-based MPC in [12] for a linear distributed system.

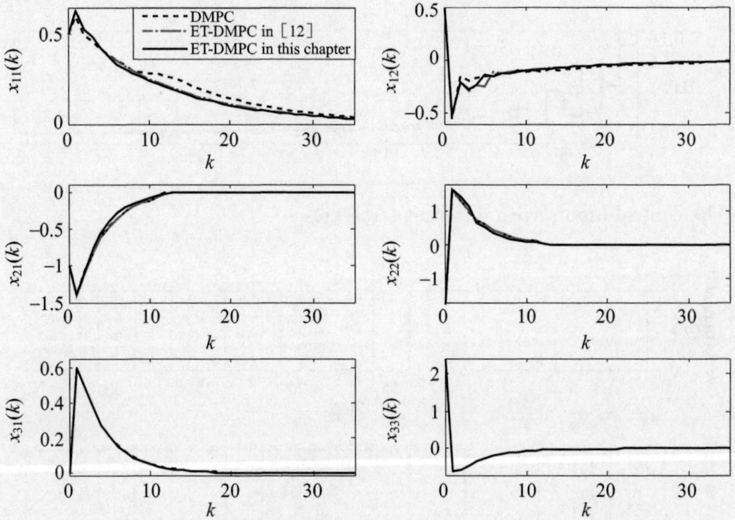

Fig. 3.2 The state trajectories with different strategies

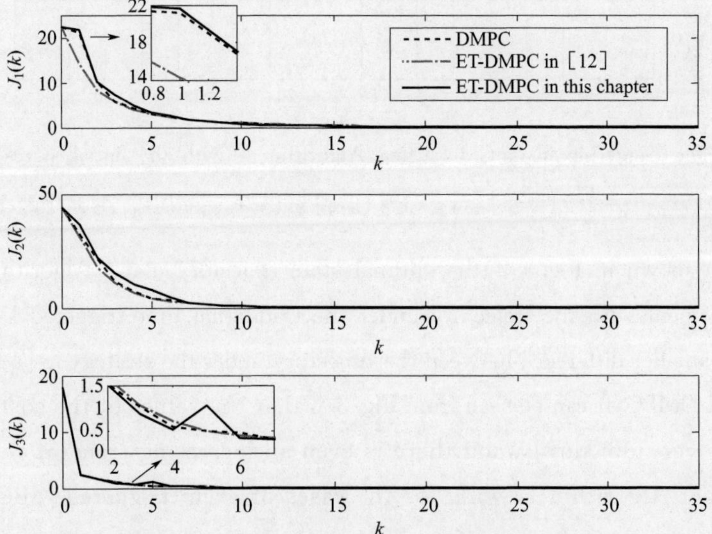

Fig. 3.3 The cost function evolutions with different strategies

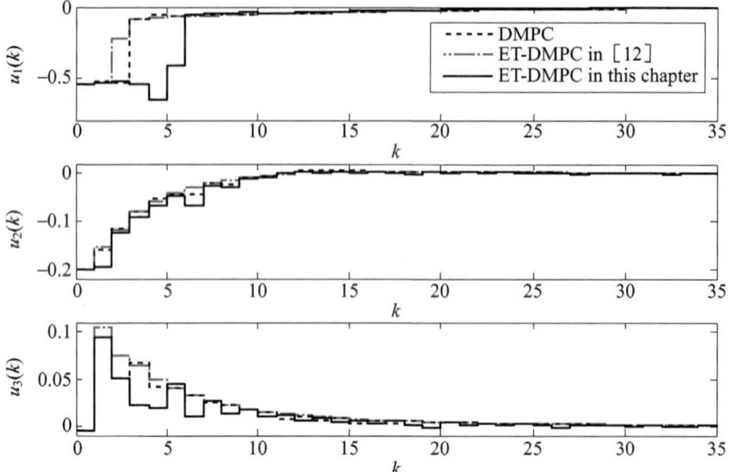

Fig. 3. 4 The control inputs with different strategies

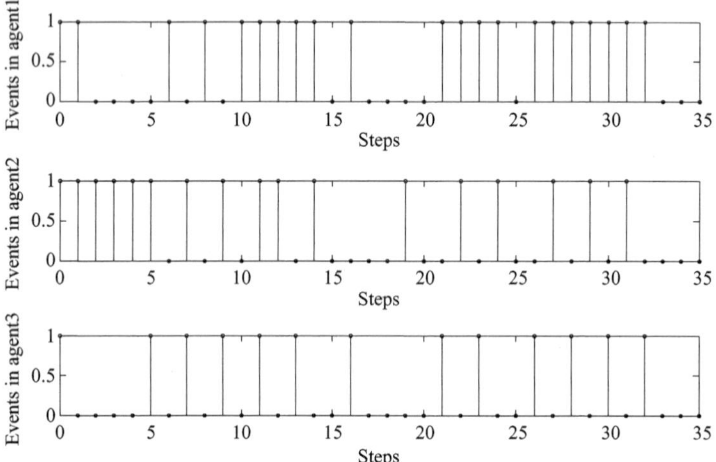

Fig. 3. 5 The triggering instants by using Algorithm 1 with the initial states $x_1(0)=[0.5,0.5]^T$, $x_2(0)=[-1,-2]^T$, $x_3(0)=[0,3]^T$ (1—Triggered, 0—Not triggered)

As is shown in Fig. 3. 2, the optimal state trajectory under Algorithm 1 is in good agreement with the trajectory under the traditional time-triggered DMPC, but there is a smaller difference between the trajectory under the strategy in [12] and the traditional DMPC. It can be seen from Fig. 3. 3 that the values of the cost function decrease over time slowly, and there is even an increasing trend at the steps 1 and 5 under Algorithm 1, while for the cases of event-triggered MPC in [12] and traditional DMPC, the performance index decreases successively.

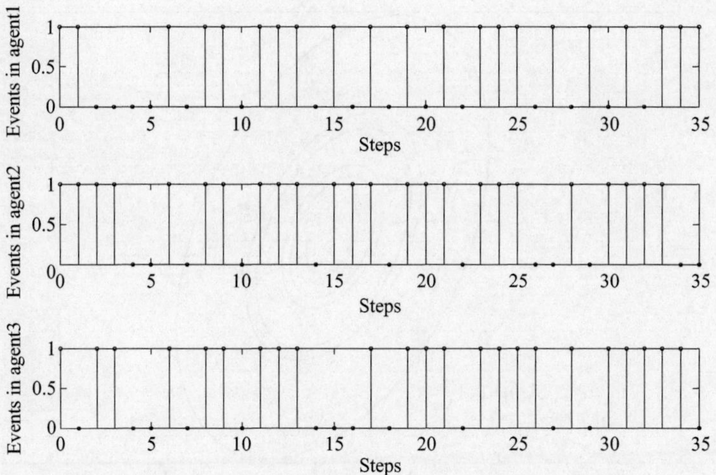

Fig. 3.6 The triggering instants by using strategy in [12] with the initial states $x_1(0)=[0.5,0.5]^T$, $x_2(0)=[-1,-2]^T$, $x_3(0)=[0,3]^T$ (1-Triggered, 0-Not triggered)

In addition, in Figs. 3.5, 3.6 and 3.8, "1" indicates that the triggering condition of the agent is satisfied at this time step; and "0" denotes that the system is not triggered, and there is no optimization problem solving and no communication between these agents. It is easy to see that the less the "1", in Figs. 3.5, 3.6 and 3.8, the lower the utilization of communication resources. Contrasting Fig. 3.5 with Fig. 3.6, we can observe that, the 1th agent, the 2th agent and 3th agent are triggered 21, 17, 13 times in 35 simulation steps with the proposed Algorithm 1, respectively, and 21, 24, 23 times in 35 simulation steps with the event-triggered MPC in [12], respectively. We can observe that, the algorithm introduced in this chapter results in a lower triggering frequency compared with that of the event-triggered MPC in [12]. What's more, in the traditional DMPC, a large amount of information is transmitted among the three agents due to 102 iterative calculations. Thus, we can obtain that the proposed algorithm saves the consumption on computation and communication.

Based on the analysis above, we illustrate that the method in this chapter achieves a trade-off between resource usage and the control performance of the system.

Then, to verify the performance of terminal set expansion method, we set $x_1(0)=[-1,-3]^T$, $x_2(0)=[1,-2]^T$, $x_3(0)=[0,-3]^T$, where the initial

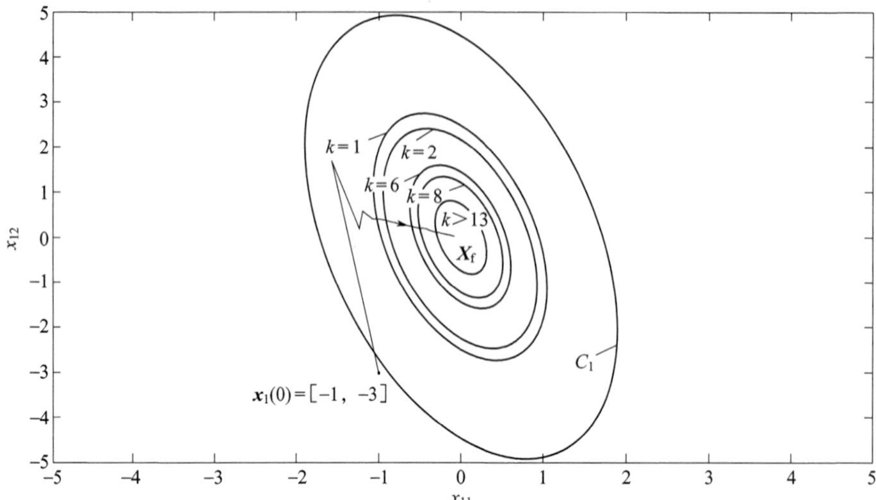

Fig. 3. 7 The adapting terminal set of the 1th agent with Algorithm 1

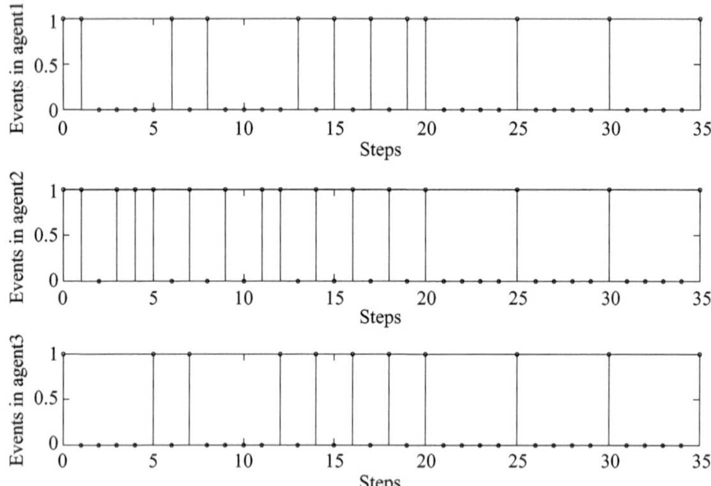

Fig. 3. 8 The triggering instants by using Algorithm 1 with the initial states $\boldsymbol{x}_1(0)=[-1,-3]^{\mathrm{T}}$, $\boldsymbol{x}_2(0)=[1,-2]^{\mathrm{T}}$, $\boldsymbol{x}_3(0)=[0,-3]^{\mathrm{T}}$ (1-Triggered, 0-Not triggered)

states are infeasible since the terminal constraints of 1th agent is violated. Solving Problem 3. 2, we obtain the minimal and optimal sets at time instants 1, 2, 6 and 8 as shown in Fig. 3. 7, which are converging to the desired terminal set $\boldsymbol{X}_f$ over time and the states of the agents are also steered to the steady-state. In addition, as shown in Fig. 3. 8, the triggering condition(3. 12)or

(3.13) is always unsatisfied as the system gradually converges to the desired states, and it tends to update periodically, i. e. the control period equals the predictive horizon $N=5$.

Remark 3.5 In Fig. 3.7, "C_1" denotes the control invariant set Φ_1 of the 1th agent, "$\boldsymbol{X}_f$" is the desired terminal set, and other ellipses are the optimal terminal sets obtained by solving Problem 3.2 at time $k=1, 2, 6$ and 8 when Problem 3.1 is infeasible. After time step 13, the terminal set remains $\boldsymbol{X}_f$.

3.6 Conclusion

The DMPC problem of NCSs with event-triggered communication to reduce unnecessary communication among nodes is considered in this chapter. The energy consumption in communication between sensors and controllers, as well as neighboring channels is reduced. The triggering condition is derived based on the Lyapunov stability analysis, and the feasibility is guaranteed by adjusting the terminal set dynamically among the algorithm implementation. Finally, the effectiveness of the proposed strategy is validated by the simulation results.

References

1. Scattolini, R. (2009). Architectures for distributed and hierarchical model predictive control-a review. *Journal of Process Control*, *19*(5), 723-731.
2. Maestre, J. M., De La Pena, D. M., Camacho, E. F., & Alamo, T. (2011). Distributed model predictive control based on agent negotiation. *Journal of Process Control*, *21*(5), 685-697.
3. Anastasi, G., Conti, M., Di Francesco, M., & Passarella, A. (2009). Energy conservation in wireless sensor networks: A survey. *Ad Hoc Networks*, *7*(3), 537-568.
4. Heemels, W. P. M. H., & Donkers, M. C. F. (2013). Model-based periodic event-triggered control for linear systems. *Automatica*, *49*(3), 698-711.
5. Heemels, W. H., Donkers, M. C. F., & Teel, A. R. (2012). Periodic event-triggered control for linear systems. *IEEE Transactions on Automatic Control*, *58*(4), 847-861.
6. Wen, G., Chen, M. Z., & Yu, X. (2015). Event-triggered master-slave synchronization with sampled-data communication. *IEEE Transactions on Circuits and Systems Ⅱ: Express Briefs*, *63*(3), 304-308.
7. Guo, G., Ding, L., & Han, Q. L. (2014). A distributed event-triggered transmission strategy for

sampled-data consensus of multi-agent systems. *Automatica*, *50*(5),1489-1496.

8. Gommans, T. M. P., & Heemels, W. P. M. H. (2015). Resource-aware MPC for constrained nonlinear systems: A self-triggered control approach. *Systems and Control Letters*, *79*,59-67.
9. Mi, X., & Li, S. (2016). Event-triggered MPC design for distributed systems with network communications. *IEEE/CAA Journal of Automatica Sinica*, *5*(1),240-250.
10. Pluymers, B., Suykens, J. A., & De Moor, B. (2005). Min-max feedback MPC using a timevarying terminal constraint set and comments on "Efficient robust constrained model predictive control with a time-varying terminal constraint set". *Systems and Control Letters*, *54* (12), 1143-1148.
11. Eqtami, A., Dimarogonas, D. V., & Kyriakopoulos, K. J. (2011). Event-triggered strategies for decentralized model predictive controllers. *IFAC Proceedings Volumes*, *44*(1),10068-10073.
12. Hashimoto, K., Adachi, S., & Dimarogonas, D. V. (2015). Distributed aperiodic model predictive control for multi-agent systems. *IET Control Theory and Applications*, *9*(1),10-20.

Chapter 4
Dynamic Event-Triggered DMPC of Networked Systems

4.1 Introduction

The DMPC of NCSs with event-triggered computation and event-triggered communication are designed in Chaps. 2 and 3, respectively. In this chapter, the trade-off between computation and communication consumption and control performance is realized by the design of a dynamic event-triggered mechanism.

In the literature on the event-triggered predictive control for NCSs, the strategy to realize the trade-off between resource usage and control performance has been an important research interest. In [1], an event-triggered DMPC is developed in a periodic fashion, and the bound of period is designed to ensure the feasibility and stability of the system. An iterative DMPC with event-triggered communication was presented in [2], where the triggering condition is related to the cost function and the bounded sub-optimality is ensured by a distributed stopping criterion. A triggering condition in [3] is designed considering the information of neighbors, and the event-based DMPC was solved in an event-triggered fashion while the information among subsystems was transformed periodically. To further save the communication costs, the triggering condition involving only the information of the subsystem itself was deduced in [4], and then the design of the event-triggered and self-triggered DMPC for the collaborative control of nonlinear NCSs is presented. Both the computation and communication resources were reduced in such control framework. and the sta-

Y. Zou and S. Li, *Distributed Cooperative Model Predictive Control of Networked Systems*, https://doi.org/10.1007/978-981-19-6084-0_4

bility and feasibility were also guaranteed. However, the control performance and convergence rate may be affected by the uncertainties caused by neighbors information.

In this chapter, a dynamic event-triggered DMPC approach with asynchronous coordination for NCSs is reported, which realizes the trade-off between computation and communication consumption and control performance. First, an event-triggered DMPC with the triggering condition related to the information of the neighbors is introduced. In this scheme, the local optimization problem computation and the communication between the subsystems occur only at the triggering time instants. To reduce computation and communication consumption further, a dynamic variable considering the influence of neighboring subsystems is employed in the design of the triggering condition and a dynamic event-triggering strategy is introduced. We show that a greater inter-execution time can be realized with the dynamic event triggering. The feasibility and closed-loop stability for the two triggering strategies are then analyzed, respectively. The main results of this chapter have been published in [5].

The organization of this chapter is as follows. Section 4. 2 states the event-triggered DMPC optimization problem for each subsystem. In Sections 4. 3 and 4. 4, the event-triggering condition and dynamic event-triggering condition are derived, respectively. The event-triggered DMPC algorithm is proposed in Section 4. 5. Furthermore, the recursive feasibility and closed-loop stability are analyzed in Section 4. 6. Section 4. 7 verifies the effectiveness of the proposed strategy with simulations. Finally, the conclusion is given in Section 4. 8.

Notation: Throughout this book, $\mathbb{R}^n$ denotes the real n dimensional Euclidean space; $\mathbb{N}$ is the collection of all natural numbers; $\boldsymbol{I}_{n\times n}$ is the identity matrix with $n\times n$ dimension. The superscript T is matrix transposition. Given a positive definite matrix Q and a column vector $\boldsymbol{x}=[x_1,\ldots,x_n]^{\mathrm{T}}$, $\bar{\lambda}(\boldsymbol{Q})$ and $\underline{\lambda}(\boldsymbol{Q})$ denote the maximum eigenvalue and minimum eigenvalue of $\boldsymbol{Q}$, respectively; $\|\boldsymbol{Q}\|=\sqrt{\bar{\lambda}(\boldsymbol{Q}^{\mathrm{T}}\boldsymbol{Q})}$ is the Euclidean norm of $\boldsymbol{Q}$; $\|\boldsymbol{x}\|_{\infty}=\max(|x_1|,\ldots,|x_n|)$ is the infinity norm of $\boldsymbol{x}$; $\|\boldsymbol{x}\|=\sqrt{\boldsymbol{x}^{\mathrm{T}}\boldsymbol{x}}$ and $\|\boldsymbol{x}\|_Q=\sqrt{\boldsymbol{x}^{\mathrm{T}}\boldsymbol{Q}\boldsymbol{x}}$ stand for the Euclidean norm and $\boldsymbol{Q}$-weighted norm of x, respectively. $\mathcal{G}=(\mathcal{V},\mathcal{E})$ denotes a graph for modeling network topology, where $\mathcal{V}=\{1,2,\ldots,M\}$ is the set of nodes (subsystems) and $\mathcal{E} \subset \mathcal{V} \times \mathcal{V}$ is the set of edges (i,j) with $i,j \in \mathcal{V}$. k_i^d ($d \in \mathbb{N}$) is the d-th triggering instant for subsystem $\mathcal{A}_i$; $k_j^{(d)_i}$ is the triggering instant closest to the time k_i^d for subsystem $\mathcal{A}_j$.

4.2 Optimization Problem Formulation

Consider a NCSs with subsystem $\mathcal{A}_i$ $(i=1,\dots,M)$ described as

$$\boldsymbol{x}_i(k+1) = \boldsymbol{A}_i \boldsymbol{x}_i(k) + \boldsymbol{B}_i \boldsymbol{u}_i(k) + \boldsymbol{w}_i(k) \tag{4.1}$$

where $\boldsymbol{x}_i(k) \in \mathbb{R}^{n_i}$, $\boldsymbol{u}_i(k) \in \mathbb{R}^{m_i}$ and $\boldsymbol{w}_i(k) \in \mathbb{R}^{n_i}$ are state, control input and disturbance, respectively; $\boldsymbol{A}_i$ and $\boldsymbol{B}_i$ are known constant matrices. Here, the constraint on control input is given by $\boldsymbol{u}_i(k) \in \boldsymbol{U}_i$, where $\boldsymbol{U}_i$ is a compact set containing the origin. The disturbance $\boldsymbol{w}_i(k)$ is bounded, i.e. $\|\boldsymbol{w}_i(k)\| \leqslant \overline{\boldsymbol{w}}_i$. A simple and undirected graph $\mathcal{G}$ is employed to describe the network topology of the M subsystems: if $(i,j) \in \mathcal{E}$, then $\mathcal{A}_i$ and $\mathcal{A}_j$ are neighboring subsystems, which can communicate with each other. Denote $\mathcal{N}_i$ as the collection of neighbors for subsystem $\mathcal{A}_i$.

The related nominal model of subsystem $\mathcal{A}_i$ of system (4.1) is defined as:

$$\tilde{\boldsymbol{x}}_i(k+1) = \boldsymbol{A}_i \tilde{\boldsymbol{x}}_i(k) + \boldsymbol{B}_i \boldsymbol{u}_i(k) \tag{4.2}$$

In this work, the event-triggered DMPC optimization problem is solved and information is transmitted to neighbors only when the event triggering condition is satisfied. Before proceeding, we first made the following assumption.

Assumption 4.1 For the nominal model (4.2), assume that (d_i, u_{d_i}) is the desired equilibrium point. Given positive definite symmetric matrices $\boldsymbol{Q}_i$ and $\boldsymbol{R}_i$, there exist a constant $\varepsilon_i > 0$, a positive definite symmetric matrix $\boldsymbol{P}_i$ and a local state feedback control law $\boldsymbol{K}_i(\tilde{\boldsymbol{x}}_i - d_i) + \boldsymbol{u}_{d_i} \in \boldsymbol{U}_i$ such that $(\boldsymbol{A}_i + \boldsymbol{B}_i \boldsymbol{K}_i)^{\mathrm{T}} \boldsymbol{P}_i (\boldsymbol{A}_i + \boldsymbol{B}_i \boldsymbol{K}_i) - \boldsymbol{P}_i + \overline{\boldsymbol{Q}}_i \leqslant 0$ for all $\tilde{\boldsymbol{x}}_i \in \phi_i = \{\tilde{\boldsymbol{x}}_i \in \mathbb{R}^{n_i} : \|\tilde{\boldsymbol{x}}_i - d_i\|_{\boldsymbol{P}_i}^2 \leqslant \varepsilon_i^2\}$, where $\overline{\boldsymbol{Q}}_i = \boldsymbol{Q}_i + \boldsymbol{K}_i^{\mathrm{T}} \boldsymbol{R}_i \boldsymbol{K}_i$.

We then give the event-triggered DMPC optimization problem at time k_i^d for subsystem $\mathcal{A}_i$ as follows:

Problem 4.1

$$\begin{aligned}&\min_{u_i(k_i^d+l\mid k_i^d)} J_i(k_i^d)\\&=\sum_{l=0}^{N-1}[\|\tilde{\boldsymbol{x}}_i(k_i^d+l\mid k_i^d)-d_i\|_{\boldsymbol{Q}_i}^2+\|\boldsymbol{u}_i(k_i^d+l\mid k_i^d)\|_{\boldsymbol{R}_i}^2\\&+\sum_{j\in\mathcal{N}_i}\|\tilde{\boldsymbol{x}}_i(k_i^d+l\mid k_i^d)-\hat{\boldsymbol{x}}_j(k_i^d+l\mid k_i^d)+\boldsymbol{d}_{ij}\|_{\boldsymbol{Q}_{ij}}^2]\\&+\|\tilde{\boldsymbol{x}}_i(k_i^d+N\mid k_i^d)-\boldsymbol{d}_i\|_{\boldsymbol{P}_i}^2\end{aligned} \tag{4.3}$$

subject to:

$$\tilde{\boldsymbol{x}}_i(k_i^d+l+1\mid k_i^d)=\boldsymbol{A}_i\tilde{\boldsymbol{x}}_i(k_i^d+l\mid k_i^d)+\boldsymbol{B}_i\boldsymbol{u}_i(k_i^d+l\mid k_i^d) \tag{4.4}$$

$$\boldsymbol{u}_i(k_i^d+l\mid k_i^d)\in\boldsymbol{U}_i,l=0,\dots,N-1 \tag{4.5}$$

$$\|\tilde{\boldsymbol{x}}_i(k_i^d+l\mid k_i^d)-\boldsymbol{d}_i\|_{\boldsymbol{P}_i}\leqslant\frac{N\theta_i\varepsilon_i}{l},l=1,\dots,N \tag{4.6}$$

where $\tilde{\boldsymbol{x}}_i(k_i^d+l\mid k_i^d)$ is the predicted state at time k_i^d+l based on the information at time k_i^d; $\boldsymbol{u}_i(k_i^d+l\mid k_i^d)$ is the control input predicted at time k_i^d; $\hat{\boldsymbol{x}}_j(k_i^d+l\mid k_i^d)$ is the estimated state of neighboring subsystem $\mathcal{A}_j$. N denotes prediction horizon; $\boldsymbol{Q}_i$ and $\boldsymbol{R}_i$ are positive definite weighting matrices; $\boldsymbol{Q}_{ij}$ is the coordination matrix between subsystem $\mathcal{A}_i$ and neighboring subsystem $\mathcal{A}_j$; $\boldsymbol{P}_i$ is the terminal weighting matrix, which satisfies Assumption 4.1. $\boldsymbol{d}_{ij}$ is the desired distance between subsystems $\mathcal{A}_i$ and $\mathcal{A}_j$, that is, for any two neighboring subsystems $\mathcal{A}_i$ and $\mathcal{A}_j$, we have $\boldsymbol{d}_i+\boldsymbol{d}_{ij}=\boldsymbol{d}_j$. The parameters $0<\theta_i<1$ and $\varepsilon_i>0$. Since the information exchange is in an asynchronous manner, $\hat{\boldsymbol{x}}_j(k_i^d+l\mid k_i^d)$ is generated based on the latest information received from subsystem $\mathcal{A}_j$, i.e. $\hat{\boldsymbol{x}}_j(k_i^d+l\mid k_i^d)=\tilde{\boldsymbol{x}}_j^*(k_i^d+l\mid k_j^{(d)_i})$, where $\tilde{\boldsymbol{x}}_j^*(k_i^d+l\mid k_j^{(d)_i})$ is the optimal predicted state of time $k_j^{(d)_i}$.

Remark 4.1 Note that the constraint (4.6) describes a decreasing upper bound on the $\boldsymbol{P}_i$-weighted norm of the difference between the predicted states and the expected state, which is employed to guarantee the robustness of the control in presence of disturbances. We subsequently prove that the convergence of the cost function can be guaranteed against disturbances with the help of (4.6).

4.3 Event-Triggering Condition

Assume that Problem 4.1 is solved at time k_i^d, then we can obtain the optimal control input $\boldsymbol{u}_i^*(k_i^d+l\,|\,k_i^d)$, the corresponding predicted state $\tilde{\boldsymbol{x}}_i^*(k_i^d+l\,|\,k_i^d)$ and cost function $J_i^*(k_i^d)$. Define k_i^{d+1} as the next triggering instant, then the candidate control sequence $\bar{u}_i(k_i^d+l\,|\,k_i^d+m)$ with $k_i^d+m\in(k_i^d, k_i^{d+1}](1\leqslant m\leqslant N-1)$ can be constructed as follows:

$$\bar{\boldsymbol{u}}_i(k_i^d+l\,|\,k_i^d+m)=\begin{cases}\boldsymbol{u}_i^*(k_i^d+l\,|\,k_i^d), l=m,\ldots,N-1\\ \boldsymbol{K}_i[\bar{\boldsymbol{x}}_i(k_i^d+l\,|\,k_i^d+m)-\boldsymbol{d}_i], l=N,\ldots,m+N-1\end{cases} \tag{4.7}$$

where $\bar{\boldsymbol{x}}_i(k_i^d+l\,|\,k_i^d+m)$ is a predicted state at time k_i^d+l based on the control sequence in (4.7). To formulate the uncertainties, a norm of difference between the current actual state $\boldsymbol{x}_i(k_i^d+m)$ and the predicted state $\tilde{\boldsymbol{x}}_i^*(k_i^d+m\,|\,k_i^d)$ calculated at the latest triggering instant is defined, i.e. $\boldsymbol{e}_i(k_i^d+m)=\|\boldsymbol{x}_i(k_i^d+m)-\tilde{\boldsymbol{x}}_i^*(k_i^d+m\,|\,k_i^d)\|$. Let $\bar{J}_i(k_i^d+m)$ present the cost function under candidate control sequence(4.7), we then consider the difference $\Delta J_i(k_i^d+m)$ between $\bar{J}_i(k_i^d+m)$ and $\bar{J}_i(k_i^d+m-1)$ as follows.

Theorem 4.1 *For subsystem $\mathcal{A}_i$ in (4.1), suppose that Assumption 4.1 is satisfied. If the event-triggering condition is designed as (4.8):*

$$\begin{aligned}&\Pi_2^i(k_i^d+m)\boldsymbol{e}_i^2(k_i^d+m)+\Pi_3^i(k_i^d+m)\boldsymbol{e}_i(k_i^d+m)\\ &>\sigma_i[\|\boldsymbol{x}_i(k_i^d+m-1)-\boldsymbol{d}_i\|_{\boldsymbol{Q}_i}^2+\|\boldsymbol{u}_i^*(k_i^d+m-1\,|\,k_i^d)\|_{\boldsymbol{R}_i}^2-\Pi_1^{ij}(k_i^d+m)]\end{aligned} \tag{4.8}$$

and

$$\bar{\boldsymbol{w}}_i\leqslant\frac{(\|\boldsymbol{A}_i\|-1)(1-\theta_i)\varepsilon_i}{\sqrt{\bar{\lambda}(\boldsymbol{P}_i)}\,\|\boldsymbol{A}_i\|(\|\boldsymbol{A}_i\|^{N-1}-1)} \tag{4.9}$$

$$\frac{\underline{\lambda}(\boldsymbol{Q}_i)}{\bar{\lambda}(\boldsymbol{P}_i)}\varepsilon_i^2-\hat{\Xi}_3^{ij}-\hat{\Xi}_4^i>0 \tag{4.10}$$

then we have $\Delta J_i(k_i^d+m)<0$, where

$$\Pi_2^i(k_i^d+m)=\bar{\lambda}(\boldsymbol{Q}_i)\sum_{l=m}^{N-1}\|\boldsymbol{A}_i\|^{2(l-m)}+\bar{\lambda}(\boldsymbol{P}_i)\|\boldsymbol{A}_i\|^{2(N-m)}$$

$$\Pi_1^{ij}(k_i^d+m)=\Xi_3^{ij}(k_i^d+m)+\Xi_4^i(k_i^d+m)$$

$$\Pi_3^i(k_i^d+m)=\frac{2\bar{\lambda}(\boldsymbol{Q}_i)N\theta_i\varepsilon_i}{\sqrt{\underline{\lambda}(\boldsymbol{P}_i)}}\sum_{l=m}^{N-1}\frac{\|\boldsymbol{A}_i\|^{l-m}}{l}+2\sqrt{\bar{\lambda}(\boldsymbol{P}_i)}\,\theta_i\varepsilon_i\|\boldsymbol{A}_i\|^{N-m}$$

$$
\begin{aligned}
&-\sum_{l=m}^{m+N-2}[\|\bar{\boldsymbol{x}}_i(k_i^d+l\,|\,k_i^d+m-1)-\boldsymbol{d}_i\|_{\boldsymbol{Q}_i}^2\\
&+\|\bar{\boldsymbol{u}}_i(k_i^d+l\,|\,k_i^d+m-1)\|_{\boldsymbol{R}_i}^2]
\end{aligned}
$$

$$
\begin{aligned}
\Xi_3^{ij}(k_i^d+m)=&\sum_{j\in\mathcal{N}_i}\left[\sqrt{\frac{\bar{\lambda}(\boldsymbol{Q}_{ij})}{\underline{\lambda}(\boldsymbol{P}_i)}}\|\boldsymbol{x}_i(k_i^d+m)-\boldsymbol{d}_i\|_{\boldsymbol{P}_i}\right.\\
&\left.+\|\tilde{\boldsymbol{x}}_j^*(k_i^d+m\,|\,k_j^{(d+m)_i})-\boldsymbol{d}_j\|_{\boldsymbol{Q}_{ij}}\right]^2\\
&+\sum_{l=m+1}^{m+N-1}\sum_{j\in\mathcal{N}_i}\left[\sqrt{\frac{\bar{\lambda}(\boldsymbol{Q}_{ij})}{\underline{\lambda}(\boldsymbol{P}_i)}}\times\frac{N\theta_i\varepsilon_i}{l-m}+\|\tilde{\boldsymbol{x}}_j^*(k_i^d+l\,|\,k_j^{(d+m)_i})-\boldsymbol{d}_j\|_{\boldsymbol{Q}_{ij}}\right]^2
\end{aligned}
$$

$$
\begin{aligned}
\Xi_4^i(k_i^d+m)=&\sum_{l=m}^{N-1}[\|\tilde{\boldsymbol{x}}_i^*(k_i^d+l\,|\,k_i^d)-\boldsymbol{d}_i\|_{\boldsymbol{Q}_i}^2+\|\boldsymbol{u}_i^*(k_i^d+l\,|\,k_i^d)\|_{\boldsymbol{R}_i}^2]\\
&+\|\tilde{\boldsymbol{x}}_i^*(k_i^d+N\,|\,k_i^d)-\boldsymbol{d}_i\|_{\boldsymbol{P}_i}^2-\sum_{l=m}^{m+N-2}[\|\bar{\boldsymbol{x}}_i(k_i^d+l\,|\,k_i^d+m-1)-\boldsymbol{d}_i\|_{\boldsymbol{Q}_i}^2\\
&+\|\bar{\boldsymbol{u}}_i(k_i^d+l\,|\,k_i^d+m-1)\|_{\boldsymbol{R}_i}^2]\\
&-\|\bar{\boldsymbol{x}}_i(k_i^d+m+N-1\,|\,k_i^d+m-1)-\boldsymbol{d}_i\|_{\boldsymbol{P}_i}^2
\end{aligned}
$$

$$
\begin{aligned}
\hat{\Xi}_3^{ij}=&\sum_{j\in\mathcal{N}_i}\left[\sqrt{\frac{\bar{\lambda}(\boldsymbol{Q}_{ij})}{\underline{\lambda}(\boldsymbol{P}_i)}}(N\theta_i\varepsilon_i+\varepsilon_i-\theta_i\varepsilon_i)+\sqrt{\frac{\bar{\lambda}(\boldsymbol{Q}_{ij})}{\underline{\lambda}(\boldsymbol{P}_j)}}N\theta_j\varepsilon_j\right]^2\\
&+\sum_{l=1}^{N-1}\sum_{j\in\mathcal{N}_i}\left[\sqrt{\frac{\bar{\lambda}(\boldsymbol{Q}_{ij})}{\underline{\lambda}(\boldsymbol{P}_i)}}\times\frac{N\theta_i\varepsilon_i}{l}+\sqrt{\frac{\bar{\lambda}(\boldsymbol{Q}_{ij})}{\underline{\lambda}(\boldsymbol{P}_j)}}\times\frac{N\theta_j\varepsilon_j}{l+1}\right]^2
\end{aligned}
$$

$$
\begin{aligned}
\hat{\Xi}_4^i=&\left[\frac{(\|\boldsymbol{A}_i\|^{N-2}-1)\bar{\boldsymbol{w}}_i}{\|\boldsymbol{A}_i\|-1}\right]^2\left[\bar{\lambda}(\boldsymbol{Q}_i)\sum_{l=2}^{N-1}\|\boldsymbol{A}_i\|^{2(l-1)}+\bar{\lambda}(\boldsymbol{P}_i)\|\boldsymbol{A}_i\|^{2(N-1)}\right]\\
&+\frac{2(\|\boldsymbol{A}_i\|^{N-2}-1)N\theta_i\varepsilon_i\bar{\boldsymbol{w}}_i}{\|\boldsymbol{A}_i\|-1}\frac{\bar{\lambda}(\boldsymbol{Q}_i)}{\sqrt{\underline{\lambda}(\boldsymbol{P}_i)}}\times\sum_{l=2}^{N-1}\frac{\|\boldsymbol{A}_i\|^{l-1}}{l-1}\\
&+\frac{\|\boldsymbol{A}_i\|^{N-1}}{2}\sqrt{\bar{\lambda}(\boldsymbol{P}_i)}]
\end{aligned}
$$

Proof Denote $\Delta J_i(k_i^d+m)=\bar{J}_i(k_i^d+m)-\bar{J}_i(k_i^d+m-1)$. By subtracting and adding $\sum_{l=m}^{N-1}[\|\tilde{\boldsymbol{x}}_i^*(k_i^d+l\,|\,k_i^d)-\boldsymbol{d}_i\|_{\boldsymbol{Q}_i}^2+\|\boldsymbol{u}_i^*(k_i^d+l\,|\,k_i^d)\|_{\boldsymbol{R}_i}^2]+\|\tilde{\boldsymbol{x}}_i^*(k_i^d+N\,|\,k_i^d)-\boldsymbol{d}_i\|_{\boldsymbol{P}_i}^2$ in $\Delta J_i(k_i^d+m)$, we have

$$
\begin{aligned}
&\Delta J_i(k_i^d+m)\\
&\leqslant-\|\boldsymbol{x}_i(k_i^d+m-1)-\boldsymbol{d}_i\|_{\boldsymbol{Q}_i}^2-\|\boldsymbol{u}_i^*(k_i^d+m-1\,|\,k_i^d)\|_{\boldsymbol{R}_i}^2\\
&+\Xi_1^i(k_i^d+m)+\Xi_2^i(k_i^d+m)+\Xi_3^{ij}(k_i^d+m)+\Xi_4^i(k_i^d+m)
\end{aligned}
\tag{4.11}
$$

where $\Xi_1^i(k_i^d+m)=\sum_{l=m}^{N-1}[\|\bar{\boldsymbol{x}}_i(k_i^d+l|k_i^d+m)-\boldsymbol{d}_i\|_{\boldsymbol{Q}_i}^2-\|\tilde{\boldsymbol{x}}_i^*(k_i^d+l|k_i^d)-\boldsymbol{d}_i\|_{\boldsymbol{Q}_i}^2]$, $\Xi_2^i(k_i^d+m)=\|\bar{\boldsymbol{x}}_i(k_i^d+m+N|k_i^d+m)-\boldsymbol{d}_i\|_{\boldsymbol{P}_i}^2-\|\tilde{\boldsymbol{x}}_i^*(k_i^d+N|k_i^d)-\boldsymbol{d}_i\|_{\boldsymbol{P}_i}^2+\sum_{l=N}^{m+N-1}\|\bar{\boldsymbol{x}}_i(k_i^d+l|k_i^d+m)-\boldsymbol{d}_i\|_{\boldsymbol{Q}_i}^2$, $\Xi_3^{ij}(k_i^d+m)=\sum_{l=m}^{m+N-1}\sum_{j\in\mathcal{N}_i}\|\bar{\boldsymbol{x}}_i(k_i^d+l|k_i^d+m)-\hat{\boldsymbol{x}}_j(k_i^d+l|k_i^d+m)+\boldsymbol{d}_{ij}\|_{\boldsymbol{Q}_{ij}}^2$. Note from (4.7) that $\bar{\boldsymbol{u}}_i(k_i^d+l|k_i^d+m)=\boldsymbol{u}_i^*(k_i^d+l|k_i^d)(l=m,\dots,N-1)$, we have $\|\bar{\boldsymbol{x}}_i(k_i^d+l|k_i^d+m)-\tilde{\boldsymbol{x}}_i^*(k_i^d+l|k_i^d)\|\leqslant\|\boldsymbol{A}_i\|^{l-m}\boldsymbol{e}_i(k_i^d+m)$. Thus, $\Xi_1^i(k_i^d+m)$ is rewritten as

$$\Xi_1^i(k_i^d+m)\leqslant\frac{2\bar{\lambda}(\boldsymbol{Q}_i)N\theta_i\varepsilon_i}{\sqrt{\underline{\lambda}(\boldsymbol{P}_i)}}\sum_{l=m}^{N-1}\frac{\|\boldsymbol{A}_i\|^{l-m}}{l}\boldsymbol{e}_i(k_i^d+m)$$
$$+\bar{\lambda}(\boldsymbol{Q}_i)\sum_{l=m}^{N-1}\|\boldsymbol{A}_i\|^{2(l-m)}\boldsymbol{e}_i^2(k_i^d+m) \tag{4.12}$$

By subtracting and adding $\sum_{l=N}^{m+N-1}\|\bar{\boldsymbol{x}}_i(k_i^d+l|k_i^d+m)-\boldsymbol{d}_i\|_{\boldsymbol{P}_i}^2$ in $\Xi_2^i(k_i^d+m)$, we can obtain

$$\Xi_2^i(k_i^d+m)\leqslant 2\sqrt{\bar{\lambda}(\boldsymbol{P}_i)}\,\theta_i\varepsilon_i\|\boldsymbol{A}_i\|^{N-m}\boldsymbol{e}_i(k_i^d+m)$$
$$+\bar{\lambda}(\boldsymbol{P}_i)\|\boldsymbol{A}_i\|^{2(N-m)}\boldsymbol{e}_i^2(k_i^d+m) \tag{4.13}$$

Since $\hat{\boldsymbol{x}}_j(k_i^d+l|k_i^d+m)=\tilde{\boldsymbol{x}}_j^*(k_i^d+l|k_j^{(d+m)_i})$, it can be obtained that

$$\Xi_3^{ij}(k_i^d+m)\leqslant\sum_{l=m}^{m+N-1}\sum_{j\in\mathcal{N}_i}[\|\bar{\boldsymbol{x}}_i(k_i^d+l|k_i^d+m)-\boldsymbol{d}_i\|_{\boldsymbol{Q}_{ij}}$$
$$+\|\tilde{\boldsymbol{x}}_j^*(k_i^d+l|k_j^{(d+m)_i}-\boldsymbol{d}_j)\|_{\boldsymbol{Q}_{ij}}]^2$$
$$\leqslant\bar{\Xi}_3^{ij}(k_i^d+m) \tag{4.14}$$

Substituting (4.12)-(4.14) into (4.11) yields that $\Delta J_i(k_i^d+m)\leqslant-\|\boldsymbol{x}_i(k_i^d+m-1)-\boldsymbol{d}_i\|_{\boldsymbol{Q}_i}^2-\|\boldsymbol{u}_i^*(k_i^d+m-1|k_i^d)\|_{\boldsymbol{R}_i}^2+\Pi_1^{ij}(k_i^d+m)+\Pi_2^i(k_i^d+m)\boldsymbol{e}_i^2(k_i^d+m)+\Pi_3^i(k_i^d+m)\boldsymbol{e}_i(k_i^d+m)$. As the control input $\boldsymbol{u}_i(k_i^d+l)=\boldsymbol{u}_i^*(k_i^d+l|k_i^d)(l=0,\dots,m-1)$ is applied in the interval $[k_i^d, k_i^d+m)$, the prediction error is bounded by

$$\boldsymbol{e}_i(k_i^d+m)=\Big\|\sum_{l=1}^{m}\boldsymbol{A}_i^{m-l}\boldsymbol{w}_i(k_i^d+l-1)\Big\|\leqslant\frac{\|\boldsymbol{A}_i\|^m-1}{\|\boldsymbol{A}_i\|-1}\bar{\boldsymbol{w}}_i \tag{4.15}$$

Thus, it follows that $\|\bar{\boldsymbol{x}}_i(k_i^d+l|k_i^d+m)-\boldsymbol{d}_i\|_{\boldsymbol{P}_i}-\|\tilde{\boldsymbol{x}}_i^*(k_i^d+l|k_i^d)-\boldsymbol{d}_i\|_{\boldsymbol{P}_i}\leqslant$

$\frac{\sqrt{\bar{\lambda}(\boldsymbol{P}_i)}(\|\boldsymbol{A}_i\|^l-\|\boldsymbol{A}_i\|^{l-m})}{\|\boldsymbol{A}_i\|-1}\overline{w}_i$. Based on condition (4.9), we have $\|\boldsymbol{x}_i(k_i^d+m)-\boldsymbol{d}_i\|_{\boldsymbol{P}_i}\leqslant N\theta_i\varepsilon_i+\varepsilon_i-\theta_i\varepsilon_i$. Noting $k_i^d+l-k_j^{(d+m)_i}\geqslant l-m+1$, we have $\Xi_3^{ij}(k_i^d+m)\leqslant\hat{\Xi}_3^{ij}$. Moreover, it is easily shown that $\Xi_4^i(k_i^d+m)\leqslant\hat{\Xi}_4^i$. Since $\boldsymbol{x}_i(k_i^d+m-1)\notin\phi_i$, it can be obtained that $\|\boldsymbol{x}_i(k_i^d+m-1)-\boldsymbol{d}_i\|_{\boldsymbol{Q}_i}^2>\frac{\underline{\lambda}(\boldsymbol{Q}_i)}{\bar{\lambda}(\boldsymbol{P}_i)}\varepsilon_i^2$. With condition (4.10), we have $\|\boldsymbol{x}_i(k_i^d+m-1)-\boldsymbol{d}_i\|_{\boldsymbol{Q}_i}^2+\|\boldsymbol{u}_i^*(k_i^d+m-1|k_i^d)\|_{\boldsymbol{R}_i}^2-\Pi_1^{ij}(k_i^d+m)>\frac{\underline{\lambda}(\boldsymbol{Q}_i)}{\bar{\lambda}(\boldsymbol{P}_i)}\varepsilon_i^2-\hat{\Xi}_3^{ij}-\hat{\Xi}_4^i>0$. If event-triggering condition (4.8) is not satisfied, we have

$$\begin{aligned}\Delta J_i(k_i^d+m)<(\sigma_i-1)[&\|\boldsymbol{x}_i(k_i^d+m-1)-\boldsymbol{d}_i\|_{\boldsymbol{Q}_i}^2\\&+\|\boldsymbol{u}_i^*(k_i^d+m-1|k_i^d)\|_{\boldsymbol{R}_i}^2-\Pi_1^{ij}(k_i^d+m)]<0\end{aligned}$$

If event-triggering condition (4.8) is satisfied, the triggering instant is updated by $k_i^d=k_i^d+m$ and the prediction error is set to $\boldsymbol{e}_i(k_i^d+m)=\|\boldsymbol{x}_i(k_i^d+m)-\tilde{\boldsymbol{x}}_i^*(k_i^d+m\mid k_i^d+m)\|=0$, which results in $\Delta J_i(k_i^d+m)\leqslant-[\|\boldsymbol{x}_i(k_i^d+m-1)-\boldsymbol{d}_i\|_{\boldsymbol{Q}_i}^2+\|\boldsymbol{u}_i^*(k_i^d+m-1|k_i^d)\|_{\boldsymbol{R}_i}^2-\Pi_1^{ij}(k_i^d+m)]<0$. Based on the discussion above, it follows that $\Delta J_i(k_i^d+m)<0$ with event-triggering condition (4.8). □

Remark 4.2 Parameter m is set to $1\leqslant m\leqslant N-1$ with the consideration of the implementation of event-triggered DMPC. That is, if condition (4.8) is not satisfied during horizon N, Problem 4.1 would be solved at time instant k_i^d+N.

Remark 4.3 In Theorem 4.1, we derive the triggering condition (4.8) by introducing the prediction difference between the actual state and the predicted one. The difference may be large due to the disturbances and further degrades the overall control performance. To ensure the closed-loop robustness, the control law is updated and the information is exchanged when the triggering condition is satisfied. Note that the triggering frequency may increase due to the existence of disturbances.

4.4 Dynamic Event-Triggering Condition

The convergence of each subsystem is realized, see Section 4.6, and the communication and computation resources are reduced with the event-triggering mechanism in Section 4.3. However, it is worth noting that, not all triggering instants generated by condition (4.8) require subsystems to transmit information and solve Problem 4.1. That is, the control performance may not be degraded, even if some control actions are not implemented at some triggering instants. Thus, to further reduce the conservatism and lower the communication and computation burdens, a dynamic variable $\eta_i(k_i^d+m)$ is employed in the design of the event triggering condition:

$$\begin{aligned}&\Pi_2^i(k_i^d+m)\boldsymbol{e}_i^2(k_i^d+m)+\Pi_3^i(k_i^d+m)\boldsymbol{e}_i(k_i^d+m)\\&\quad>\sigma_i[\|\boldsymbol{x}_i(k_i^d+m-1)-\boldsymbol{d}_i\|_{\boldsymbol{Q}_i}^2+\|\boldsymbol{u}_i^*(k_i^d+m-1|k_i^d)\|_{\boldsymbol{R}_i}^2\\&\quad-\Pi_1^{ij}(k_i^d+m)]+\rho_i\eta_i(k_i^d+m)\end{aligned}\tag{4.16}$$

with $\rho_i>0$ and $\eta_i(k_i^d+m)=\delta_i\eta_i(k_i^d+m-1)-\Pi_2^i(k_i^d+m)\boldsymbol{e}_i^2(k_i^d+m)-\Pi_3^i(k_i^d+m)\boldsymbol{e}_i(k_i^d+m)+\sigma_i[\|\boldsymbol{x}_i(k_i^d+m-1)-\boldsymbol{d}_i\|_{\boldsymbol{Q}_i}^2+\|\boldsymbol{u}_i^*(k_i^d+m-1|k_i^d)\|_{\boldsymbol{R}_i}^2-\Pi_1^{ij}(k_i^d+m)]$ $(0<\delta_i<1)$. Note that condition (4.8) can be regarded as a specific case of the dynamic event-triggering condition (4.16) when ρ_i approaches 0.

The advantage of introducing the dynamic variable $\eta_i(k_i^d+m)$ is shown in corollary 4.1.

Corollary 4.1 *For subsystem $\mathcal{A}_i$, the inter-execution time under dynamic event-triggering condition (4.16) is greater than that under event-triggering condition (4.8). In other words, computation and communication resources can be further reduced with condition (4.16).*

Proof Assume that the most recent triggering instant is k_i^d. Let k_i^{d+1} and $\hat{k}_i^{d+1}$ denote the next triggering instant obtained by event-triggering condition (4.8) and dynamic event-triggering condition (4.16), respectively. We then compare the size of these two time instants. Suppose that $k_i^{d+1}\geqslant\hat{k}_i^{d+1}$. When using event-triggering condition (4.8), $\hat{k}_i^{d+1}$ is in the interval (k_i^d, k_i^{d+1}) so that we can get $\Pi_2^i(\hat{k}_i^{d+1})\boldsymbol{e}_i^2(\hat{k}_i^{d+1})+\Pi_3^i(\hat{k}_i^{d+1})\boldsymbol{e}_i(\hat{k}_i^{d+1})\leqslant\sigma_i[\|\boldsymbol{x}_i(\hat{k}_i^{d+1}-1)-\boldsymbol{d}_i\|_{\boldsymbol{Q}_i}^2+\|\boldsymbol{u}_i^*(\hat{k}_i^{d+1}-1|k_i^d)\|_{\boldsymbol{R}_i}^2-\Pi_1^{ij}(\hat{k}_i^{d+1})]$. With dynamic event-triggering condition (4.16), at triggering instant $\hat{k}_i^{d+1}$, we have $\Pi_2^i(\hat{k}_i^{d+1})\boldsymbol{e}_i^2(\hat{k}_i^{d+1})+\Pi_3^i(\hat{k}_i^{d+1})$

$\boldsymbol{e}_i(\hat{k}_i^{d+1}) > \sigma_i[\|\boldsymbol{x}_i(\hat{k}_i^{d+1}-1)-\boldsymbol{d}_i\|_{\boldsymbol{Q}_i}^2 + \|\boldsymbol{u}_i^*(\hat{k}_i^{d+1}-1|k_i^d)\|_{\boldsymbol{R}_i}^2 - \Pi_1^{ij}(\hat{k}_i^{d+1})]$, which contradicts with the former inequality. Then it can be concluded that the inter-execution time under dynamic event-triggering condition (4.16) is greater than that under event-triggering condition (4.8). □

4.5 Dynamic Event-Triggered DMPC Algorithm

The dual-mode control scheme is adopted in the implementation of the event-triggered DMPC. Before the subsystems enter their invariant sets, they communicate with each other and solve Problem 4.1 in an event-triggered manner. Once the subsystems enter the invariant set, they adopt a local state feedback control law instead of solving Problem 4.1 and communicating with each other. The proposed dynamic event-triggered distributed predictive control algorithm is shown in the sequel.

Algorithm 1 Dynamic Event-triggered Asynchronous DMPC

for $k=0,1,2,\ldots$ **do**

  if $k=0$ **then**

    if $\boldsymbol{x}_i(0) \in \phi_i$ **then**

      the state feedback control law $\boldsymbol{u}_i(0)=\boldsymbol{K}_i\boldsymbol{x}_i(0)$ is applied.

    else

      let $\hat{\boldsymbol{x}}_j(l|0)=0(l=0,\ldots,N-1)$. Solve Problem 4.1 based on $\boldsymbol{x}_i(0)$ to yield the optimal control sequence $\boldsymbol{u}_i^*(0|0),\ldots,\boldsymbol{u}_i^*(N-1|0)$. Apply the first control input $\boldsymbol{u}_i^*(0|0)$ to $\mathcal{A}_i$. Subsystem $\mathcal{A}_i$ transmits the predicted states $\tilde{\boldsymbol{x}}_i^*(1|0),\ldots,\tilde{\boldsymbol{x}}_i^*(N|0)$ to neighboring subsystem $\mathcal{A}_j$.

    end if

  else

    if $\boldsymbol{x}_i(k) \in \boldsymbol{\phi}_i$ **then**

      the state feedback control law $\boldsymbol{u}_i(k)=\boldsymbol{K}_i\boldsymbol{x}_i(k)$ is applied.

    else

      if the event-triggering condition (4.8) or (4.16) is satisfied **then**

        Update the triggering instant by $k_i^d=k$. Solve Problem 4.1 based on $\boldsymbol{x}_i(k_i^d)$ and received predicted states $\tilde{\boldsymbol{x}}_j^*(k_i^d|k_j^{(d)_i}),\ldots,\tilde{\boldsymbol{x}}_j^*(k_i^d+N-1|k_j^{(d)_i})$ to yield the optimal control sequence $\boldsymbol{u}_i^*(k_i^d|k_i^d),\ldots,\boldsymbol{u}_i^*(k_i^d+N-1|k_i^d)$. Apply the first control input $\boldsymbol{u}_i^*(k_i^d|k_i^d)$ to $\mathcal{A}_i$. Subsystem $\mathcal{A}_i$ transmits the predicted states $\tilde{\boldsymbol{x}}_i^*(k_i^d+1|k_i^d),\ldots,\tilde{\boldsymbol{x}}_i^*(k_i^d+N|k_i^d)$ to neighboring subsystem $\mathcal{A}_j$.

      else

        Problem 4.1 is not solved and candidate control sequence (4.7) is applied. Moreover: no information is transmitted from subsystem $\mathcal{A}_i$ to neighbouring subsystem $\mathcal{A}_j$.

      end if

```
        end if
    end if
end for
```

4.6 Performance Analysis

We first analyze the convergence of subsystem $\mathcal{A}_i$ with event-triggering condition as in (4.8).

Theorem 4.2 *Given subsystem $\mathcal{A}_i$ in (4.1), candidate control inputs (4.7) and event-triggering condition (4.8). If conditions (4.9)-(4.10) hold and parameter θ_i satisfies the following condition:*

$$\max\left\{\sqrt{1-\frac{\underline{\lambda}(\overline{\boldsymbol{Q}}_i)}{\lambda(\boldsymbol{P}_i)}},1-\frac{1}{1+\|\boldsymbol{A}_i\|^{N-2}(N-1)}\right\}\leqslant\theta_i \tag{4.17}$$

then subsystem $\mathcal{A}_i$ will converge to its invariant set $\boldsymbol{\phi}_i$.

Proof The proof of Theorem 4.2 consists of two parts. We first prove the recursive feasibility of the candidate control sequence (4.7), i.e. (4.7) satisfies constraints (4.5) and (4.6).

(i) $\|\overline{\boldsymbol{x}}_i(k_i^d+l\,|\,k_i^d+m)-\boldsymbol{d}_i\|_{\boldsymbol{P}_i}\leqslant\frac{N\theta_i\varepsilon_i}{l-m}(l=m+1,\ldots,m+N)$.

Let $l=N$ and according to condition (4.9), we can get $\|\overline{\boldsymbol{x}}_i(k_i^d+N\,|\,k_i^d+m)-\boldsymbol{d}_i\|_{\boldsymbol{P}_i}\leqslant\varepsilon_i$, which implies $\overline{\boldsymbol{x}}_i(k_i^d+l\,|\,k_i^d+m)\in\boldsymbol{\phi}_i$ with $N+1\leqslant l\leqslant m+N$. Thus, we have $\|\overline{\boldsymbol{x}}_i(k_i^d+l\,|\,k_i^d+m)-\boldsymbol{d}_i\|_{\boldsymbol{P}_i}^2-\|\overline{\boldsymbol{x}}_i(k_i^d+l-1\,|\,k_i^d+m)-\boldsymbol{d}_i\|_{\boldsymbol{P}_i}^2\leqslant-\|\overline{\boldsymbol{x}}_i(k_i^d+l-1\,|\,k_i^d+m)-\boldsymbol{d}_i\|_{\boldsymbol{Q}_i}^2$. Using condition (4.17), we can see that $\|\overline{\boldsymbol{x}}_i(k_i^d+m+N\,|\,k_i^d+m)-\boldsymbol{d}_i\|_{\boldsymbol{P}_i}^2\leqslant\left[1-\frac{\underline{\lambda}(\overline{\boldsymbol{Q}}_i)}{\lambda(\boldsymbol{P}_i)}\right](\varepsilon_i)^2\leqslant(\theta_i\varepsilon_i)^2$. To make $\|\overline{\boldsymbol{x}}_i(k_i^d+l\,|\,k_i^d+m)-\boldsymbol{d}_i\|_{\boldsymbol{P}_i}\leqslant\frac{N\theta_i\varepsilon_i}{l-m}(l=m+1,\ldots,m+N-1)$ hold, we have to prove $\frac{\sqrt{\lambda(\boldsymbol{P}_i)}(\|\boldsymbol{A}_i\|^{l}-\|\boldsymbol{A}_i\|^{l-m})}{\|\boldsymbol{A}_i\|-1}\overline{\boldsymbol{w}}_i\leqslant\frac{mN\theta_i\varepsilon_i}{l(l-m)}$. According to condition (4.9), we have $\frac{\sqrt{\bar{\lambda}(\boldsymbol{P}_i)}(\|\boldsymbol{A}_i\|^l-\|\boldsymbol{A}_i\|^{l-m})}{\|\boldsymbol{A}_i\|-1}\overline{w}_i\leqslant\|\boldsymbol{A}_i\|^{N-2}(1-\theta_i)$ ε_i. Noting $l\leqslant m+N-1$ and $m\geqslant 1$, it is obtained that $\frac{mN\theta_i\varepsilon_i}{l(l-m)}\geqslant\frac{\theta_i\varepsilon_i}{N-1}$. Furthermore, with condition (4.17), we can get $\|\boldsymbol{A}_i\|^{N-2}(1-\theta_i)\varepsilon_i\leqslant\frac{\theta_i\varepsilon_i}{N-1}$. As a result, we have $\|\overline{\boldsymbol{x}}_i(k_i^d+l\,|\,k_i^d+m)-\boldsymbol{d}_i\|_{\boldsymbol{P}_i}\leqslant\frac{N\theta_i\varepsilon_i}{l-m}(l=m+1,\ldots,m+N)$.

(ii) $\overline{\boldsymbol{u}}_i(k_i^d+l\,|\,k_i^d+m)\in\boldsymbol{U}_i(l=m,\ldots,m+N-1)$.

Based on the definition of control sequence (4.7), we have $\overline{\boldsymbol{u}}_i(k_i^d+l\,|\,k_i^d+m)=\boldsymbol{u}_i^*(k_i^d+l\,|\,k_i^d)\in\boldsymbol{U}_i(l=m,\ldots,N-1)$. Since $\overline{\boldsymbol{x}}_i(k_i^d+l\,|\,k_i^d+m)\in\boldsymbol{\phi}_i(l=N,\ldots,m+N-1)$, we can obtain $\overline{\boldsymbol{u}}_i(k_i^d+l\,|\,k_i^d+m)=\boldsymbol{K}_i[\overline{\boldsymbol{x}}_i(k_i^d+l\,|\,k_i^d+m)-\boldsymbol{d}_i)]\in\boldsymbol{U}_i(l=N,\ldots,m+N-1)$. From the above, we have $\overline{\boldsymbol{u}}_i(k_i^d+l\,|\,k_i^d+m)\in\boldsymbol{U}_i(l=m,\ldots,m+N-1)$. It can be deduce that the candidate control sequence $\overline{\boldsymbol{u}}_i(k_i^d+l\,|\,k_i^d+m)$ is recursively feasible.

We then show the convergence of subsystem $\mathcal{A}_i$ with event-triggering condition(4.8). It can be seen from Theorem 4.1 that $\Delta J_i(k_i^d+m)<0$ under event-triggering condition (4.8), thus we can obtain that subsystem $\mathcal{A}_i$ will enter its invariant set. □

Remark 4.4 From Theorem 4.2, it can be seen that the upper bound of disturbance $\overline{w}_i$ may affect the feasibility and convergence of the control. As shown in (4.9), the upper bound of disturbances should not be too large, especially for unstable systems.

We then give the analysis of the convergence of subsystem $\mathcal{A}_i$ with dynamic event-triggering condition (4.16). Before proceeding, we first give the following lemma.

Lemma 4.1 *For a given initial value $\eta_i(k_i^d)>0$, the value of the dynamic variable $\eta_i(k_i^d+m)$ remains positive in the DMPC implementation.*

Proof Once the dynamic event-triggering condition (4.16) is satisfied, the difference $\boldsymbol{e}_i(k_i^d+m)$ will be renewed to 0. Thus, $\eta_i(k_i^d+m-1)=\frac{1}{\delta_i}\{\eta_i(k_i^d+m)-$

$\sigma_i[\|\boldsymbol{x}_i(k_i^d+m-1)-\boldsymbol{d}_i\|_{\boldsymbol{Q}_i}^2+\|\boldsymbol{u}_i^*(k_i^d+m-1|k_i^d)\|_{\boldsymbol{R}_i}^2-\Pi_1^{ij}(k_i^d+m)]\} \leqslant \frac{1}{\delta_i}\eta_i(k_i^d+m)$. If dynamic event-triggering condition (4.16) is not satisfied, we have $\eta_i(k_i^d+m-1)=\frac{1}{\delta_i}\{\eta_i(k_i^d+m)+\Pi_2^i(k_i^d+m)\boldsymbol{e}_i^2(k_i^d+m)+\Pi_3^i(k_i^d+m)\boldsymbol{e}_i(k_i^d+m)-\sigma_i[\|\boldsymbol{x}_i(k_i^d+m-1)-\boldsymbol{d}_i\|_{\boldsymbol{Q}_i}^2+\|\boldsymbol{u}_i^*(k_i^d+m-1|k_i^d)\|_{\boldsymbol{R}_i}^2-\Pi_1^{ij}(k_i^d+m)]\} \leqslant \frac{1+\rho_i}{\delta_i}\eta_i(k_i^d+m)$. Therefore, given an initial value $\eta_i(k_i^d)>0$, one can conclude that $\eta_i(k_i^d+m)>0$. □

Theorem 4.3 *If conditions (4.9), (4.10) and (4.17) hold, subsystem $\mathcal{A}_i$ will enter its invariant set $\boldsymbol{\phi}_i$ under dynamic event-triggering condition (4.16).*

Proof For subsystem $\mathcal{A}_i$, the Lyapunov function is chosen as $\boldsymbol{W}_i[\boldsymbol{x}_i(k_i^d+m),\eta_i(k_i^d+m)]=\overline{J}_i[\boldsymbol{x}_i(k_i^d+m)]+\eta_i(k_i^d+m)$. From Lemma 4.1, it is obvious that $\boldsymbol{W}_i[\boldsymbol{x}_i(k_i^d+m),\eta_i(k_i^d+m)]>0$ for an initial value $\eta_i(k_i^d)>0$. The difference $\Delta\boldsymbol{W}_i[\boldsymbol{x}_i(k_i^d+m)]$ between $\boldsymbol{W}_i[\boldsymbol{x}_i(k_i^d+m),\eta_i(k_i^d+m)]$ and $\boldsymbol{W}_i[\boldsymbol{x}_i(k_i^d+m-1),\eta_i(k_i^d+m-1)]$ is given by $\Delta\boldsymbol{W}_i[\boldsymbol{x}_i(k_i^d+m),\eta_i(k_i^d+m)]=\Delta J_i(k_i^d+m)+\eta_i(k_i^d+m)-\eta_i(k_i^d+m-1) \leqslant (\sigma_i-1)[\|\boldsymbol{x}_i(k_i^d+m-1)-\boldsymbol{d}_i\|_{\boldsymbol{Q}_i}^2+\|\boldsymbol{u}_i^*(k_i^d+m-1|k_i^d)\|_{\boldsymbol{R}_i}^2-\Pi_1^{ij}(k_i^d+m)]+(\delta_i-1)\eta_i(k_i^d+m-1)$. According to Lemma 4.1, we have $\Delta\boldsymbol{W}_i[\boldsymbol{x}_i(k_i^d+m),\eta_i(k_i^d+m)]<0$, which means that subsystem $\mathcal{A}_i$ will converge to its invariant set under dynamic event-triggering condition (4.16). □

Remark 4.5 The trade-off between resource usage and control performance can be realized by the design of parameters σ_i and ρ_i. That is, the smaller the values of σ_i and ρ_i, the easier it is to satisfy the dynamic event-triggering condition, which will also increase the consumption of the communication and computation. The larger the values of σ_i and ρ_i, the stricter the dynamic event-triggering condition will be and the lower the triggering frequency, which, however, may degrade the control performance. Note that, with the increasing of ρ_i, the DMPC controller may not be triggered during the prediction horizon N, and Problem 4.1 will be solved at time k_i^d+N.

4.7 Example

In this section, we consider a multi-agent formation example [7,8], in which the discretized system is described as:

$$\begin{cases} p_i^x(k+1)=p_i^x(k)+Tv_i^x(k)+T^2/2m_if_i^x(k) \\ v_i^x(k+1)=v_i^x(k)+T/m_if_i^x(k) \\ p_i^y(k+1)=p_i^y(k)+Tv_i^y(k)+T^2/2m_if_i^y(k) \\ v_i^y(k+1)=v_i^y(k)+T/m_if_i^y(k) \end{cases}$$

where $p_i^x(k)$ and $p_i^y(k)$ are the horizonal position and vertical position, respectively; $v_i^x(k)$ is the horizonal velocity and $v_i^y(k)$ is the vertical velocity; $f_i^x(k)$ is the horizonal force and $f_i^y(k)$ is the vertical force. $T=0.1$s is the discretized period and $m_i=0.1$kg is the mass of agent i. The control input satisfies $\|u_i(k)\|_\infty \leqslant 1$.

The parameters in Problem 4.1 are chosen as $N=3$, $\boldsymbol{Q}_i=\boldsymbol{I}_{4\times 4}$, $\boldsymbol{R}_i=5\boldsymbol{I}_{2\times 2}$, $\varepsilon_i=0.2$. The desired distances are set as $\boldsymbol{d}_{12}=[0.1,0,-0.2,0]$, $\boldsymbol{d}_{23}=[-0.1,0,-0.2,0]$, $\boldsymbol{d}_{34}=[-0.1,0,0.2,0]$, $\boldsymbol{d}_{41}=[0.1,0,0.2,0]$. The terminal weighting matrix is designed as $\boldsymbol{P}_i=[4.5862, 2.2913, 0, 0; 2.2913, 4.2693, 0, 0; 0, 0, 4.5862, 2.2913; 0, 0, 2.2913, 4.2693]$ and the state feedback control law is given as $\boldsymbol{K}_i=[-0.3424, -0.4309, 0, 0; 0, 0, -0.3424, -0.4309]$. According to Theorem 4.1, we set the upper bound of disturbances as $\overline{\boldsymbol{w}}_i=0.01$ and choose the coordination matrix as $\boldsymbol{Q}_{ij}=[0.1, 0, 0, 0; 0, 0, 0, 0; 0, 0, 0, 1, 0; 0, 0, 0, 0, 0]$. From Theorem 4.2, $\theta_i=0.7966$. The parameters σ_i, δ_i and ρ_i are given as 0.7, 0.3 and 1.

The initial conditions of Algorithm 1 are set as: $\boldsymbol{x}_1(0)=[0.4,0,0.4,0]^{\mathrm{T}}$, $\boldsymbol{x}_2(0)=[0.5,0,0.1,0]^{\mathrm{T}}$, $\boldsymbol{x}_3(0)=[0.4,0,-0.4,0]^{\mathrm{T}}$, $\boldsymbol{x}_4(0)=[0.3,0,-0.1,0]^{\mathrm{T}}$, $\eta_1(0)=\eta_2(0)=\eta_3(0)=\eta_4(0)=1$. As shown in Figs. 4.1 and 4.2, four agents converge to the desired position while keeping a desired distance. The control inputs are illustrated in Fig. 4.3. The triggering instants generated by triggering condition (4.8) and dynamic triggering condition (4.16) are presented in Figs. 4.4 and 4.5, respectively. It can be seen that the computation and communication frequency is reduced significantly under dynamic event-triggering condition (4.16). The influence of σ_i and ρ_i on the triggered frequency is illustrated in Table 4.1, which illustrates that the triggered frequency decreases with the increase of σ_i, and the triggered frequency under dynamic triggering condition (4.16) is getting closer to that under triggering condition (4.8) with the decrease of ρ_i.

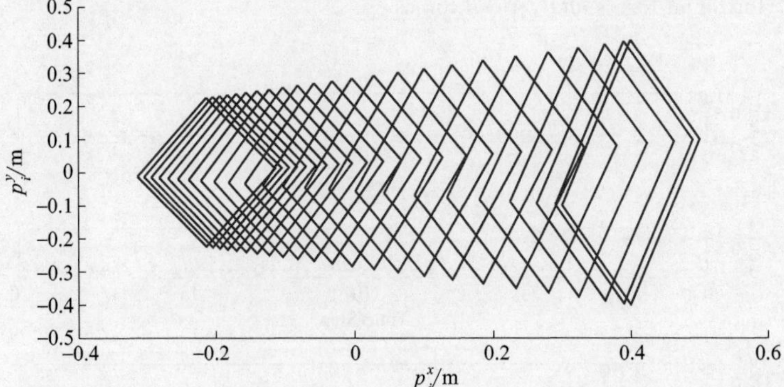

Fig. 4.1 Movement trajectories of agents

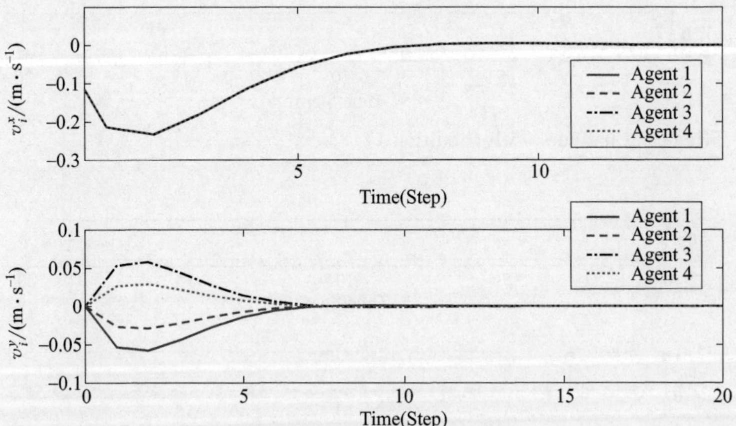

Fig. 4.2 Horizontal velocities and vertical velocities

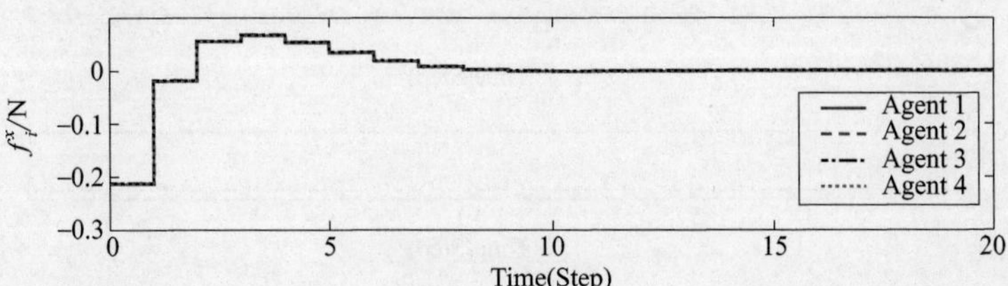

Fig. 4.3

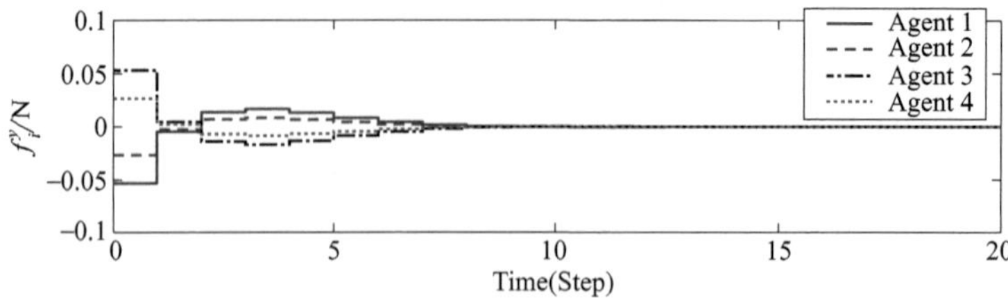

Fig. 4. 3 Horizontal forces and vertical forces

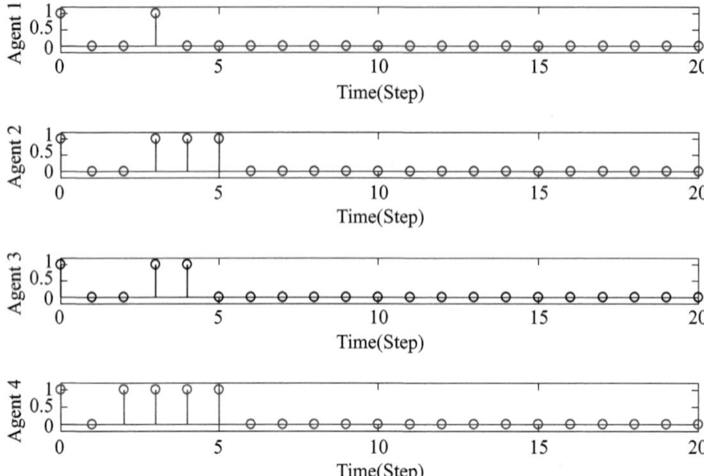

Fig. 4. 4 Triggering instants with condition (4. 8)

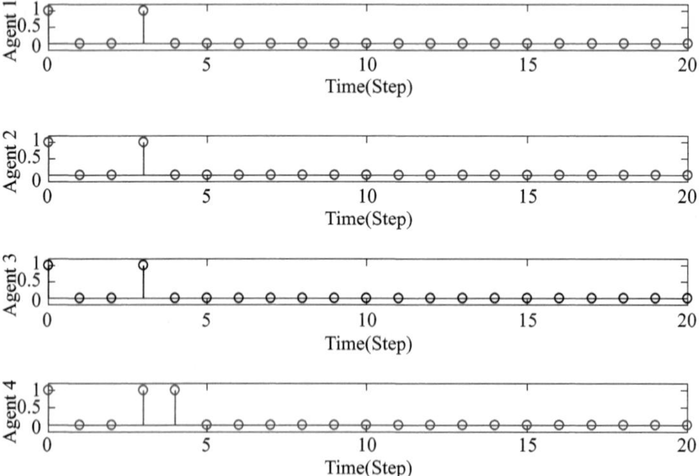

Fig. 4. 5 Triggering instants with condition (4. 16)

Table 4.1 Total triggered frequency of agents

	$\sigma_i=0.1$	$\sigma_i=0.7$
Triggered frequency of Condition(4.8)	18	14
Triggered frequency of Condition(4.16) ($\rho_i=1.0$)	13	9
Triggered frequency of Condition(4.16) ($\rho_i=0.1$)	15	12

4.8 Conclusion

In this chapter, an event-triggered DMPC scheme for uncertain linear NCSs has been proposed, where two types of triggering conditions have been designed to realize the trade-off between resource utilization and control performance. Especially, the computational and communication burden is reduced more significantly under the proposed dynamic event-triggering condition. The feasibility of the designed algorithm and stability of the closed-loop systems have been analyzed.

References

1. Li, H., Yan, W., Shi, Y., & Wang, Y. (2015). Periodic event-triggering in distributed receding horizon control of nonlinear systems. *Systems and Control Letters*, *86*, 16-23.
2. Groβ, D., & Stursberg, O. (2015). A cooperative distributed MPC algorithm with event-based communication and parallel optimization. *IEEE Transactions on Control of Network Systems*, *3*(3), 275-285.
3. Zou, Y., Su, X., & Niu, Y. (2016). Event-triggered distributed predictive control for the cooperation of multi-agent systems. *IET Control Theory and Applications*, *11*(1), 10-16.
4. Hashimoto, K., Adachi, S., & Dimarogonas, D. V. (2015). Distributed aperiodic model predictive control for multi-agent systems. *IET Control Theory and Applications*, *9*(1), 10-20.
5. Zou, Y., Su, X., Li, S., Niu, Y., & Li, D. (2019). Event-triggered distributed predictive control for asynchronous coordination of multi-agent systems. *Automatica*, *99*, 92-98.
6. Girard, A. (2014). Dynamic triggering mechanisms for event-triggered control. *IEEE Transac-*

tions on Automatic Control, *60*(7), 1992-1997.

7. Ding, B., Xie, L., & Cai, W. (2010). Distributed model predictive control for constrained linear systems. *International Journal of Robust and Nonlinear Control*, *20*(11), 1285-1298.
8. Wang, C., & Ong, C. J. (2010). Distributed model predictive control of dynamically decoupled systems with coupled cost. *Automatica*, *46*(12), 2053-2058.

Chapter 5
Mixed Time/Event-Triggered DMPC of Networked Systems

5.1 Introduction

The wired networks have matured in large-scale systems. However, it is difficult to wire in harsh environments and long-distance cabling will result in high consumption. Therefore, wireless components are introduced for parts that cannot be realized by wired components. Thus, the mixed wired-wireless network has become a development trend for large-scale systems [1-4]. For wireless sensor nodes, event-triggered communication should be employed considering the limited wireless resources. On the other hand, it is necessary to adopt a time-triggered mechanism in the wired network to guarantee the overall control performance. As a result, the time-triggered and event-triggered schemes exist simultaneously in distributed systems.

Event-triggered DMPC has been studied a lot in the existing literature [5-8]. However, in the mixed triggered networked environment, the pattern of control and communication has changed fundamentally, which can not be handled by simply extending the existing single time-triggered or event-triggered approaches. Specifically, since time-triggered and event-triggered subsystems coexist, they are not able to coordinate with each other synchronously, which may reduce the control performance or even affect the stability of the system. In addition, the recursive feasibility of DMPC and the closed-loop stability of the systems should be reconsidered.

In this chapter, a mixed time/event-triggered dual-mode DMPC strategy is de-

Y. Zou and S. Li, *Distributed Cooperative Model Predictive Control of Networked Systems*,
https://doi.org/10.1007/978-981-19-6084-0_5

veloped for large-scale systems under wired-wireless networks. The main results of this chapter have been published in [9] and the main features can be summarized as follows. First, by employing a prediction difference between the current actual state and the predicted one, the triggering condition of event-triggered subsystems is designed considering the influence of time/event-triggered communication patterns. Then a mixed triggered dual-mode DMPC algorithm is formulated, where subsystems optimize in a distributed manner and coordinate asynchronously. The presented mixed triggered DMPC achieves the trade-off between control performance and resource consumption, and the recursive feasibility and closed-loop stability are also guaranteed under the established sufficient conditions.

The organization of this chapter is as follows. In Section 5. 2, the optimization problems of event-triggered and time-triggered subsystems are formulated. The event-triggering condition is derived in Section 5. 3 and the mixed time/event-triggered dualmode DMPC algorithm is then presented. Section 5. 4 discusses the feasibility and stability of the DMPC. Section 5. 5 illustrates the effectiveness of the designed algorithm with simulations. Finally, the conclusion is presented in Section 5. 6.

Notation: Throughout this chapter, S_i^e denotes the ith event-triggered subsystem and S_j^t denotes the jth time-triggered subsystem. $\mathbb{N}$ is the collection of all natural numbers; $\mathbb{R}^n$ denotes the real n dimensional Euclidean space; $\boldsymbol{I}_{n\times n}$ is the identity matrix with $n\times n$ dimension. The superscript T is the matrix transposition. Given a positive definite matrix $\boldsymbol{Q}$ and a column vector $\boldsymbol{x}=[x_1, x_2, \ldots, x_n]^{\mathrm{T}}$, $\overline{\lambda}(\boldsymbol{Q})$ and $\underline{\lambda}(\boldsymbol{Q})$ are the maximum eigenvalue and minimum eigenvalue of $\boldsymbol{Q}$, respectively; $\|\boldsymbol{Q}\|=\sqrt{\overline{\lambda}(\boldsymbol{Q}^{\mathrm{T}}\boldsymbol{Q})}$ is the Euclidean norm of $\boldsymbol{Q}$; $\|\boldsymbol{x}\|_\infty=\max(|x_1|, |x_2|, \ldots, |x_n|)$ is the infinity norm of $\boldsymbol{x}$; $\|\boldsymbol{x}\|=\sqrt{\boldsymbol{x}^{\mathrm{T}}\boldsymbol{x}}$ and $\|\boldsymbol{x}\|_{\boldsymbol{Q}}=\sqrt{\boldsymbol{x}^{\mathrm{T}}\boldsymbol{Q}\boldsymbol{x}}$ stand for the Euclidean norm and $\boldsymbol{Q}$-weighted norm of $\boldsymbol{x}$.

5. 2 Problem Formulation

The structure of mixed time/event triggered DMPC can be illustrated in Fig. 5. 1. For each event-triggered subsystem at each time step, once the triggering condition is satisfied, the DMPC optimization problem is solved and the predicted states based on the optimal control sequence are transmitted to its neighboring subsystems. Otherwise, there is no information exchange between subsystems and the control sequence obtained at the most recent triggering instant will be employed. For those subsystems in a time-triggered fashion, the computation of DMPC optimization problem and the transmission of predicted state are carried out periodically. These two subsystems are modeled in the sequel, respectively.

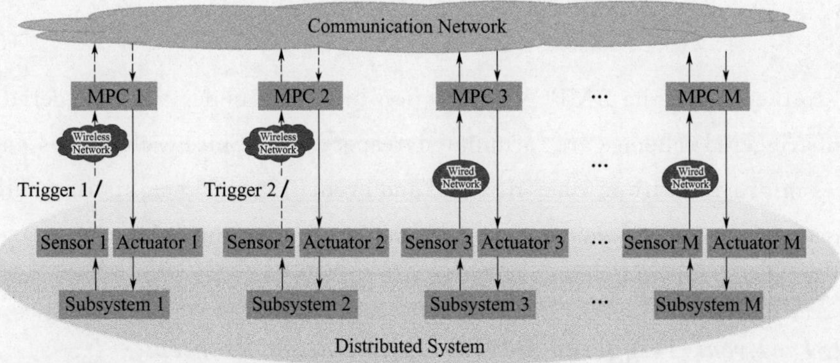

Fig. 5.1 The structure of mixed time/event-triggered DMPC: the solid lines indicate that the information is transmitted periodically, and the dotted lines indicate that the information transmission is driven by event triggers

The event-triggered subsystem $S_i^e\,(i\in\{1,\ldots,M\})$ can be formulated as:

$$\boldsymbol{x}_i^e(k+1)=\boldsymbol{A}_i^e\boldsymbol{x}_i^e(k)+\boldsymbol{B}_i^e\boldsymbol{u}_i^e(k)+\boldsymbol{w}_i^e(k) \tag{5.1}$$

where $\boldsymbol{x}_i^e(k)\in\mathbb{R}^{n_i}$, $\boldsymbol{u}_i^e(k)\in\mathbb{R}^{m_i}$ and $\boldsymbol{w}_i^e(k)\in\mathbb{R}^{n_i}$ are the state, control input and disturbance of S_i^e, respectively; $\boldsymbol{A}_i^e$ and $\boldsymbol{B}_i^e$ are system matrices. Control inputs are constrained by $\boldsymbol{u}_i^e(k)\in\boldsymbol{U}_i^e$, where $\boldsymbol{U}_i^e$ is a compact set containing the origin. The disturbance $\boldsymbol{w}_i^e(k)$ is bounded by $\|\boldsymbol{w}_i^e(k)\|\leqslant\overline{\boldsymbol{w}}_i^e$.

The corresponding nominal model of the event-triggered subsystem S_i^e can be defined as:

$$\boldsymbol{x}_i^e(k+1)=\boldsymbol{A}_i^e\boldsymbol{x}_i^e(k)+\boldsymbol{B}_i^e\boldsymbol{u}_i^e(k) \tag{5.2}$$

The time-triggered subsystem S_i^t can be formulated as:

$$x_i^t(k+1)=A_i^t x_i^t(k)+B_i^t u_i^t(k)+w_i^t(k) \tag{5.3}$$

where $x_i^t(k)\in\mathbb{R}^{n_i}$, $u_i^t(k)\in\mathbb{R}^{m_i}$ and $w_i^t(k)\in\mathbb{R}^{n_i}$ denote the state, control input and disturbance of S_i^t, respectively; A_i^t and B_i^t are system matrices. The control input is constrained by $u_i^t(k)\in U_i^t$, where U_i^t is a compact set containing the origin. The disturbance $w_i^t(k)$ is bounded, i. e. $\|w_i^t(k)\|\leqslant\overline{w}_i^t$.

The corresponding nominal model of the time-triggered subsystem S_i^t can be formulated as

$$x_i^t(k+1)=A_i^t x_i^t(k)+B_i^t u_i^t(k) \tag{5.4}$$

In the sequel, the DMPC optimization problems under time-triggered and event-triggered schemes are formulated, respectively. Since each subsystem receives information from time-triggered and event-triggered neighboring subsystems simultaneously, two types of information are considered in the design of the optimization problem.

5.2.1 Event-Triggered DMPC Optimization Problem

Problem 5.1 For event-triggered subsystem S_i^e, let $k_i^d, d\in\mathbb{N}$ indicate its dth triggering instant, and set $k_i^0=0$. The DMPC optimization problem at time k_i^d can be described as follows:

$$\begin{aligned}\min_{u_i^e(k_i^d+l\,|\,k_i^d)} J_i^e(k_i^d)=&\sum_{l=0}^{N-1}\Big[\|x_i^e(k_i^d+l\,|\,k_i^d)\|_{Q_i^e}^2+\|u_i^e(k_i^d+l\,|\,k_i^d)\|_{R_i^e}^2+\\&\sum_{j\in\mathcal{N}_i^t}\|x_i^e(k_i^d+l\,|\,k_i^d)-\hat{x}_j^t(k_i^d+l\,|\,k_i^d)\|_{Q_{ij}^{et}}^2+\\&\sum_{j\in\mathcal{N}_i^e}\|x_i^e(k_i^d+l\,|\,k_i^d)-\hat{x}_j^e(k_i^d+l\,|\,k_i^d)\|_{Q_{ij}^{ee}}^2\Big]+\\&\|x_i^e(k_i^d+N\,|\,k_i^d)\|_{P_i^e}^2\end{aligned}$$

subject to:

$$x_i^e(k_i^d+l+1\,|\,k_i^d)=A_i^e x_i^e(k_i^d+l\,|\,k_i^d)+B_i^e u_i^e(k_i^d+l\,|\,k_i^d) \tag{5.5}$$

$$\hat{x}_j^t(k_i^d+l+1\,|\,k_i^d)=A_j^t \hat{x}_j^t(k_i^d+l\,|\,k_i^d)+B_j^t \hat{u}_j^t(k_i^d+l\,|\,k_i^d) \tag{5.6}$$

$$\hat{x}_j^e(k_i^d+l+1\,|\,k_i^d)=A_j^e \hat{x}_j^e(k_i^d+l\,|\,k_i^d)+B_j^e \hat{u}_j^e(k_i^d+l\,|\,k_i^d) \tag{5.7}$$

$$u_i^e(k_i^d+l\,|\,k_i^d)\in U_i^e,\ l=0,\dots,N-1 \tag{5.8}$$

$$\|x_i^e(k_i^d+l\,|\,k_i^d)\|_{P_i^e}\leqslant\frac{N\alpha_i^e\varepsilon_i^e}{l},l=1,\ldots,N \tag{5.9}$$

where N denotes the prediction horizon ;Q_i^e and R_i^e denote the positive definite weighting matrices;P_i^e is the terminal weighting matrix, which can be designed according to the Riccatti equation, i. e. $P_i^e = (A_i^e)^T P_i^e A_i^e - (A_i^e)^T P_i^e B_i^e [R_i^e + (B_i^e)^T P_i^e B_i^e]^{-1} (B_i^e)^T P_i^e A_i^e + Q_i^e$. For a given constant $\varepsilon_i^e > 0$, $\phi_i^e = \{x_i^e \in \mathbb{R}^{n_i} : \|x_i^e\|_{P_i^e}^2 \leqslant (\varepsilon_i^e)^2\}$ indicates an invariant set, i. e. for all $x_i^e \in \phi_i^e$, we have$(A_i^e + B_i^e K_i^e) x_i^e \in \phi_i^e$ with a local state feedback control law $K_i^e x_i^e \in U_i^e$. The indices of S_i^e's neighboring subsystems is denoted as $\mathcal{N}_i^e$; the cooperation matri ces of subsystem S_i^e with neighbors S_j^t and S_j^e are represented by Q_{ij}^{et} and Q_{ij}^{ee}, respectively; and we have $0<\alpha_i^e<1$. $x_i^e(k_i^d+l\,|\,k_i^d)$ denotes the predicted state at time k_i^d+l based on the information of time k_i^d, and $u_i^e(k_i^d+l\,|\,k_i^d)$ is the predicted control input. The estimated state $\hat{x}_j^t(k_i^d+l\,|\,k_i^d)$ and the estimated control input $\hat{u}_j^t(k_i^d+l\,|\,k_i^d)$ of S_j^t are generated by S_i^e based on the information received from S_j^t; let $\hat{x}_j^t(k_i^d\,|\,k_i^d)=x_j^{t*}(k_i^d\,|\,k_i^d-1)$, where $x_j^{t*}(k_i^d\,|\,k_i^d-1)$ indicates the predicted state at time k_i^d under the optimal control sequence at time k_i^d-1; $\hat{u}_j^t(k_i^d+l\,|\,k_i^d)$ is defined by $\hat{u}_j^t(k_i^d+l\,|\,k_i^d)=u_j^{t*}(k_i^d+l\,|\,k_i^d-1)$ with $l=0,\ldots,N-2$ and $\hat{u}_j^t(k_i^d+N-1\,|\,k_i^d)=K_j^t\,\hat{x}_j^t(k_i^d+N-1\,|\,k_i^d)$, where $u_j^{t*}(k_i^d+l\,|\,k_i^d-1)$ is the optimal solution at time k_i^d-1. $\hat{x}_j^e(k_i^d+l\,|\,k_i^d)$ is the estimated state of S_j^e and define $\hat{x}_j^e(k_i^d\,|\,k_i^d)=x_j^{e*}(k_i^d\,|\,k_j^{ei})$, where k_j^{ei} is the most recent triggering instant away from k_i^d for S_j^e, i. e. $k_j^{ei}\triangleq\sup\{k_j^p\in\mathbb{N}:k_j^p<k_i^d\}$, and the superscript ei means that k_j^{ei} is associated with S_i^e; the estimated control input $\hat{u}_j^e(k_i^d+l\,|\,k_i^d)$ is defined as: for $l=0,\ldots,k_j^{ei}+N-1-k_i^d$, $\hat{u}_j^e(k_i^d+l\,|\,k_i^d)=u_j^{e*}(k_i^d+l\,|\,k_j^{ei})$ and for $l=k_j^{ei}+N-k_i^d,\ldots,N-1$, $\hat{u}_j^e(k_i^d+l\,|\,k_i^d)=K_j^e\,\hat{x}_j^e(k_i^d+l\,|\,k_i^d)$.

5.2.2 Time-Triggered DMPC Optimization Problem

Problem 5.2 For time-triggered subsystem S_i^t, the DMPC optimization problem at time k can be formulated as follows:

$$\begin{aligned}\min_{u_i^t(k+l\,|\,k)} J_i^t(k) = & \sum_{l=0}^{N-1} [\|x_i^t(k+l\,|\,k)\|_{Q_i^t}^2 + \|u_i^t(k+l\,|\,k)\|_{R_i^t}^2 \\ & + \sum_{j\in\mathcal{N}_i^t} \|x_i^t(k+l\,|\,k) - \hat{x}_j^t(k+l\,|\,k)\|_{Q_{ij}^{tt}}^2 \\ & + \sum_{j\in\mathcal{N}_i^t} \|x_i^t(k+l\,|\,k) - \hat{x}_j^e(k+l\,|\,k)\|_{Q_{ij}^{te}}^2] \\ & + \|x_i^t(k+N\,|\,k)\|_{P_i^t}^2 \end{aligned} \tag{5.10}$$

subject to:

$$x_i^t(k+l+1\,|\,k) = A_i^t x_i^t(k+l\,|\,k) + B_i^t u_i^t(k+l\,|\,k) \tag{5.11}$$

$$\hat{x}_j^t(k+l+1\,|\,k) = A_j^t \hat{x}_j^t(k+l\,|\,k) + B_j^t \hat{u}_j^t(k+l\,|\,k) \tag{5.12}$$

$$\hat{x}_j^e(k+l+1\,|\,k) = A_j^e \hat{x}_j^e(k+l\,|\,k) + B_j^e \hat{u}_j^e(k+l\,|\,k) \tag{5.13}$$

$$u_i^t(k+l\,|\,k) \in U_i^t, l=0,\ldots,N-1 \tag{5.14}$$

$$\|x_i^t(k+l\,|\,k)\|_{P_i^t} \leqslant \frac{N\alpha_i^t \varepsilon_i^t}{l}, l=1,\ldots,N \tag{5.15}$$

where Q_i^t and R_i^t are weighting matrices; the terminal weighting matrix P_i^t can be designed according to the Riccati equation, i. e. $P_i^t = (A_i^t)^T P_i^t A_i^t - (A_i^t)^T P_i^t B_i^t [R_i^t + (B_i^t)^T P_i^t B_i^t]^{-1} (B_i^t)^T P_i^t A_i^t + Q_i^t$. Similarly to the event-triggered case, $\phi_i^t = \{x_i^t \in \mathbb{R}^{n_i} : \|x_i^t\|_{P_i^t}^2 \leqslant (\varepsilon_i^t)^2, \varepsilon_i^t > 0\}$ is an invariant set for the system in(5. 4)associated with a local state feedback control law $K_i^t x_i^t \in U_i^t$; $\mathcal{N}_i^t$ is the collection of indices of S_i^t's neighboring subsystems; the cooperation matrices of subsystem S_i^t with neighbors S_j^t and S_j^e are denoted by Q_{ij}^{tt} and Q_{ij}^{te}, respectively; and we also have $0<\alpha_i^t<1$. $\hat{x}_j^t(k+l\,|\,k)$ and $\hat{u}_j^t(k+l\,|\,k)$ are the estimated state and the estimated control input of S_j^t based on the information at time k, respectively; $\hat{x}_j^t(k\,|\,k) = x_j^{t*}(k\,|\,k-1)$ and $\hat{u}_j^t(k+l\,|\,k)$ is defined by $u_j^t(k+l\,|\,k) = u_j^{t*}(k+l\,|\,k-1)$ with $l=0,\ldots,N-2$ and $\hat{u}_j^t(k+N-1\,|\,k) = K_j^t \hat{x}_j^t(k+N-1\,|\,k)$. $\hat{x}_j^e(k+l\,|\,k)$ is the estimated state of S_j^e and define $\hat{x}_j^e(k\,|\,k) = x_j^{e*}(k\,|\,k_j^{ti})$, where k_j^{ti} is the most recent triggering instant away from k for S_j^e, i. e. $k_j^{ti} \triangleq \sup\{k_j^p \in \mathbb{N} : k_j^p < k\}$, and the superscript *ti* indicates that k_j^{ti} is associated with S_i^t; the estimated control input $\hat{u}_j^e(k+$

$l|k)$ is defined as follows: for $l=0,\ldots,k_j^{ti}+N-1-k$, $\hat{\boldsymbol{u}}_j^e(k+l|k)=\boldsymbol{u}_j^{e*}(k+l|k_j^{ti})$ and for $l=k_j^{ti}+N-k,\ldots,N-1$, $\hat{\boldsymbol{u}}_j^e(k+l|k)=\boldsymbol{K}_j^e\hat{\boldsymbol{x}}_j^e(k+l|k)$.

5.3 Mixed Time/Event-Triggered Dual-Mode DMPC

In this work, we aim to design a mixed time/event-triggered DMPC to ensure the performance of the system while reducing the resource cost. Firstly, an eventtriggering condition is designed for the event-triggered subsystems, and then a mixed time/event-triggered dual-mode DMPC algorithm is presented.

5.3.1 Event-Triggering Condition

Solve Problem 5.1 at time k_i^d, then the optimal control sequence $\boldsymbol{u}_i^{e*}(k_i^d+l|k_i^d)$, the corresponding predicted state $\boldsymbol{x}_i^{e*}(k_i^d+l|k_i^d)$ and cost function $J_i^{e*}(k_i^d)$ can be obtained. Let k_i^{d+1} indicate the the next triggering instant, then the control sequence $\overline{\boldsymbol{u}}_i^e(k_i^d+l|k_i^d+m)$, $k_i^d+m\in(k_i^d,k_i^{d+1}]$ $(1\leqslant m\leqslant N-1)$ is given as:

$$\overline{\boldsymbol{u}}_i^e(k_i^d+l|k_i^d+m)=\begin{cases}\boldsymbol{u}_i^{e*}(k_i^d+l|k_i^d), l=m,\ldots,N-1\\ \boldsymbol{K}_i^e\overline{\boldsymbol{x}}_i^e(k_i^d+l|k_i^d+m), l=N,\ldots,m+N-1\end{cases} \tag{5.16}$$

where the predicted state $\overline{\boldsymbol{x}}_i^e(k_i^d+l|k_i^d+m)$ is obtained based on the corresponding control sequence in (5.16). However, there is a prediction error between the actual state and predicted one because of the existence of uncertainties. To guarantee the closed-loop stability of the subsystem, Problem 5.1 should be solved when the prediction error exceeds the tolerance. Thus, the prediction error $\delta_i(m)\triangleq\|\boldsymbol{x}_i^e(k_i^d+m)-\boldsymbol{x}_i^{e*}(k_i^d+m|k_i^d)\|$ is considered in the design of the DMPC optimization problem.

Let $\overline{J}_i^e(k_i^d+m)$ denote the cost function obtained by control sequence (5.16). We then present the difference between $\overline{J}_i^e(k_i^d+m)$ and $\overline{J}_i^e(k_i^d+m-1)$.

$$\Delta\overline{J}_i^e(k_i^d+m)$$

$$=\overline{J}_i^e(k_i^d+m)-\{\sum_{l=m}^{N-1}[\|\boldsymbol{x}_i^{e*}(k_i^d+l|k_i^d)\|_{\boldsymbol{Q}_i^e}^2+\|\boldsymbol{u}_i^{e*}(k_i^d+l|k_i^d)\|_{\boldsymbol{R}_i^e}^2]$$

$$+\|x_i^{e*}(k_i^d+N|k_i^d)\|_{P_i^e}^2\}-\bar{J}_i^e(k_i^d+m-1)+\{\sum_{l=m}^{N-1}[\|x_i^{e*}(k_i^d+l|k_i^d)\|_{Q_i^e}^2$$
$$+\|u_i^{e*}(k_i^d+l|k_i^d)\|_{R_i^e}^2]+\|x_i^{e*}(k_i^d+N|k_i^d)\|_{P_i^e}^2\} \quad (5.17)$$
$$\leqslant-\|x_i^e(k_i^d+m-1)\|_{Q_i^e}^2-\|u_i^{e*}(k_i^d+m-1|k_i^d)\|_{R_i^e}^2+\Gamma_{i1}^e(m)$$
$$+\Gamma_{i2}^e(m)+\Gamma_{i3}^e(m)+\Gamma_{i4}^e(m)+\Gamma_{i5}^e(m)$$

where

$$\Gamma_{i1}^e(m)=\sum_{l=m}^{N-1}[\|\bar{x}_i^e(k_i^d+l|k_i^d+m)\|_{Q_i^e}^2-\|x_i^{e*}(k_i^d+l|k_i^d)\|_{Q_i^e}^2],$$
$$\Gamma_{i2}^e(m)=\|\bar{x}_i^e(k_i^d+m+N|k_i^d+m)\|_{P_i^e}^2-\|x_i^{e*}(k_i^d+N|k_i^d)\|_{P_i^e}^2$$
$$+\sum_{l=N}^{m+N-1}\|\bar{x}_i^e(k_i^d+l|k_i^d+m)\|_{\bar{Q}_i^e}^2 \text{ with } \bar{Q}_i^e=Q_i^e+(K_i^e)^{\mathrm{T}}R_i^eK_i^e,$$
$$\Gamma_{i3}^e(m)=\sum_{l=m}^{m+N-1}\sum_{j\in\mathcal{N}_i^e}\|\bar{x}_i^e(k_i^d+l|k_i^d+m)-\hat{x}_j^t(k_i^d+l|k_i^d+m)\|_{Q_{ij}^{et}}^2,$$
$$\Gamma_{i4}^e(m)=\sum_{l=m}^{m+N-1}\sum_{j\in\mathcal{N}_i^e}\|\bar{x}_i^e(k_i^d+l|k_i^d+m)-\hat{x}_j^e(k_i^d+l|k_i^d+m)\|_{Q_{ij}^{ee}}^2,$$
$$\Gamma_{i5}^e(m)=\sum_{l=m}^{N-1}[\|x_i^{e*}(k_i^d+l|k_i^d)\|_{Q_i^e}^2+\|u_i^{e*}(k_i^d+l|k_i^d)\|_{R_i^e}^2]+\|x_i^{e*}(k_i^d+N|k_i^d)\|_{P_i^e}^2$$
$$-\sum_{l=m}^{m+N-2}[\|\bar{x}_i^e(k_i^d+l|k_i^d+m-1)\|_{Q_i^e}^2+\|\bar{u}_i^e(k_i^d+l|k_i^d+m-1)\|_{R_i^e}^2]$$
$$-\|\bar{x}_i^e(k_i^d+m+N-1|k_i^d+m-1)\|_{P_i^e}^2.$$

Since $\bar{u}_i^e(k_i^d+l|k_i^d+m)=u_i^{e*}(k_i^d+l|k_i^d)(l=m,\ldots,N-1)$, we have

$$\|\bar{x}_i^e(k_i^d+l|k_i^d+m)-x_i^{e*}(k_i^d+l|k_i^d)\|\leqslant\|A_i^e\|^{l-m}\delta_i(m) \quad (5.18)$$

Based on (5.18) and employing the triangle inequality, it follows

$$\Gamma_{i1}^e(m)\leqslant\frac{2\bar{\lambda}(Q_i^e)}{\sqrt{\underline{\lambda}(P_i^e)}}\sum_{l=m}^{N-1}\|x_i^{e*}(k_i^d+l|k_i^d)\|_{P_i^e}\cdot\|\bar{x}_i^e(k_i^d+l|k_i^d+m)$$
$$-x_i^{e*}(k_i^d+l|k_i^d)\|+\bar{\lambda}(Q_i^e)\sum_{l=m}^{N-1}\|\bar{x}_i^e(k_i^d+l|k_i^d+m)$$
$$-x_i^{e*}(k_i^d+l|k_i^d)\|^2\leqslant\frac{2\bar{\lambda}(Q_i^e)N\alpha_i^e\varepsilon_i^e}{\sqrt{\underline{\lambda}(P_i^e)}}\sum_{l=m}^{N-1}\frac{\|A_i^e\|^{l-m}}{l}\delta_i(m)$$
$$+\bar{\lambda}(Q_i^e)\sum_{l=m}^{N-1}\|A_i^e\|^{2(l-m)}\delta_i^2(m) \quad (5.19)$$

Moreover, we have

$$\Gamma_{i2}^{e}(m)-\sum_{l=N}^{m+N-1}\|\overline{\boldsymbol{x}}_{i}^{e}(k_{i}^{d}+l\,|\,k_{i}^{d}+m)\|_{\boldsymbol{P}_{i}^{e}}^{2}+\sum_{l=N}^{m+N-1}\|\overline{\boldsymbol{x}}_{i}^{e}(k_{i}^{d}+l\,|\,k_{i}^{d}+m)\|_{\boldsymbol{P}_{i}^{e}}^{2}$$
$$\leqslant\|\overline{\boldsymbol{x}}_{i}^{e}(k_{i}^{d}+N\,|\,k_{i}^{d}+m)\|_{\boldsymbol{P}_{i}^{e}}^{2}-\|\boldsymbol{x}_{i}^{e*}(k_{i}^{d}+N\,|\,k_{i}^{d})\|_{\boldsymbol{P}_{i}^{e}}^{2}$$
$$\leqslant 2\sqrt{\overline{\lambda}(\boldsymbol{P}_{i}^{e})}\,\alpha_{i}^{e}\varepsilon_{i}^{e}\|\boldsymbol{A}_{i}^{e}\|^{N-m}\delta_{i}(m)+\overline{\lambda}(\boldsymbol{P}_{i}^{e})\|\boldsymbol{A}_{i}^{e}\|^{2(N-m)}\delta_{i}^{2}(m) \tag{5.20}$$

Noting $\hat{\boldsymbol{x}}_{j}^{t}(k_{i}^{d}+l\,|\,k_{i}^{d}+m)=\boldsymbol{x}_{j}^{t*}(k_{i}^{d}+l\,|\,k_{i}^{d}+m-1)$ and $\hat{\boldsymbol{x}}_{j}^{e}(k_{i}^{d}+l\,|\,k_{i}^{d}+m)=\boldsymbol{x}_{j}^{e*}(k_{i}^{d}+l\,|\,k_{j}^{ei+m})(l=m,\ldots,m+N-1)$, where k_{j}^{ei+m} is the most recent triggering instant away from $k_{i}^{d}+m$, it can then be deduced that

$$\Gamma_{i3}^{e}(m)\leqslant\sum_{l=m}^{m+N-1}\sum_{j\in\mathcal{N}_{i}^{e}}\left[\|\overline{\boldsymbol{x}}_{i}^{e}(k_{i}^{d}+l\,|\,k_{i}^{d}+m)\|_{\boldsymbol{Q}_{ij}^{et}}+\|\boldsymbol{x}_{j}^{t*}(k_{i}^{d}+l\,|\,k_{i}^{d}+m-1)\|_{\boldsymbol{Q}_{ij}^{et}}\right]^{2}$$
$$\leqslant\sum_{j\in\mathcal{N}_{i}^{e}}\left[\sqrt{\frac{\overline{\lambda}(\boldsymbol{Q}_{ij}^{et})}{\underline{\lambda}(\boldsymbol{P}_{i}^{e})}}\|\boldsymbol{x}_{i}^{e}(k_{i}^{d}+m)\|_{\boldsymbol{P}_{i}^{e}}+\frac{\sqrt{\overline{\lambda}(\boldsymbol{Q}_{ij}^{et})}\,N\alpha_{j}^{t}\varepsilon_{j}^{t}}{\sqrt{\underline{\lambda}(\boldsymbol{P}_{j}^{t})}}\right]^{2}$$
$$+\sum_{l=m+1}^{m+N-1}\sum_{j\in\mathcal{N}_{i}^{e}}\left[\frac{\sqrt{\overline{\lambda}(\boldsymbol{Q}_{ij}^{et})}\,N\alpha_{i}^{e}\varepsilon_{i}^{e}}{\sqrt{\underline{\lambda}(\boldsymbol{P}_{i}^{e})}(l-m)}+\frac{\sqrt{\overline{\lambda}(\boldsymbol{Q}_{ij}^{et})}\,N\alpha_{j}^{t}\varepsilon_{j}^{t}}{\sqrt{\underline{\lambda}(\boldsymbol{P}_{j}^{t})}(l-m+1)}\right]^{2}$$
$$\triangleq\overline{\Gamma}_{i3}^{e}(m) \tag{5.21}$$

$$\Gamma_{i4}^{e}(m)\leqslant\sum_{l=m}^{m+N-1}\sum_{j\in\mathcal{N}_{i}^{e}}\left[\|\overline{\boldsymbol{x}}_{i}^{e}(k_{i}^{d}+l\,|\,k_{i}^{d}+m)\|_{\boldsymbol{Q}_{ij}^{ee}}+\|\boldsymbol{x}_{j}^{e*}(k_{i}^{d}+l\,|\,k_{j}^{ei+m})\|_{\boldsymbol{Q}_{ij}^{ee}}\right]^{2}$$
$$\leqslant\sum_{j\in\mathcal{N}_{i}^{e}}\left[\sqrt{\frac{\overline{\lambda}(\boldsymbol{Q}_{ij}^{ee})}{\underline{\lambda}(\boldsymbol{P}_{i}^{e})}}\|\boldsymbol{x}_{i}^{e}(k_{i}^{d}+m)\|_{\boldsymbol{P}_{i}^{e}}+\frac{\sqrt{\overline{\lambda}(\boldsymbol{Q}_{ij}^{ee})}\,N\alpha_{j}^{e}\varepsilon_{j}^{e}}{\sqrt{\underline{\lambda}(\boldsymbol{P}_{j}^{e})}(k_{i}^{d}+m-k_{j}^{ei+m})}\right]^{2}$$
$$+\sum_{l=m+1}^{m+N-1}\sum_{j\in\mathcal{N}_{i}^{e}}\left[\frac{\sqrt{\overline{\lambda}(\boldsymbol{Q}_{ij}^{ee})}\,N\alpha_{i}^{e}\varepsilon_{i}^{e}}{\sqrt{\underline{\lambda}(\boldsymbol{P}_{i}^{e})}(l-m)}+\frac{\sqrt{\overline{\lambda}(\boldsymbol{Q}_{ij}^{ee})}\,N\alpha_{j}^{e}\varepsilon_{j}^{e}}{\sqrt{\underline{\lambda}(\boldsymbol{P}_{j}^{e})}(k_{i}^{d}+l-k_{j}^{ei+m})}\right]^{2}$$
$$\triangleq\overline{\Gamma}_{i4}^{e}(m) \tag{5.22}$$

Substituting(5.19)-(5.22)into(5.17), it follows that

$$\Delta\overline{J}_{i}^{e}(k_{i}^{d}+m)\leqslant-\Theta_{i1}(m)+\Theta_{i2}(m)\delta_{i}^{2}(m)+\Theta_{i3}(m)\delta_{i}(m) \tag{5.23}$$

where

$$\Theta_{i1}(m)=\|\boldsymbol{x}_{i}^{e}(k_{i}^{d}+m-1)\|_{\boldsymbol{Q}_{i}^{e}}^{2}+\|\boldsymbol{u}_{i}^{e*}(k_{i}^{d}+m-1\,|\,k_{i}^{d})\|_{\boldsymbol{R}_{i}^{e}}^{2}$$
$$-\overline{\Gamma}_{i3}^{e}(m)-\overline{\Gamma}_{i4}^{e}(m)-\Gamma_{i5}^{e}(m),$$

$$\Theta_{i2}(m)=\overline{\lambda}(\boldsymbol{Q}_i^e)\sum_{l=m}^{N-1}\|\boldsymbol{A}_i^e\|^{2(l-m)}+\overline{\lambda}(\boldsymbol{P}_i^e)\|\boldsymbol{A}_i^e\|^{2(N-m)},$$

$$\Theta_{i3}(m)=\frac{2\overline{\lambda}(\boldsymbol{Q}_i^e)N\alpha_i^e\varepsilon_i^e}{\sqrt{\underline{\lambda}(\boldsymbol{P}_i^e)}}\sum_{l=m}^{N-1}\frac{\|\boldsymbol{A}_i^e\|^{l-m}}{l}+2\sqrt{\overline{\lambda}(\boldsymbol{P}_i^e)}\alpha_i^e\varepsilon_i^e\|\boldsymbol{A}_i^e\|^{N-m}.$$

Let $\Theta_{i2}(m)\delta_i^2(m)+\Theta_{i3}(m)\delta_i(m)\leqslant\sigma_i\Theta_{i1}(m)$, then we have $\Delta\overline{J}_i^e(k_i^d+m)<0$. Thus, the event-triggering condition at time k_i^d+m can be derived as:

$$\Theta_{i2}(m)\delta_i^2(m)+\Theta_{i3}(m)\delta_i(m)>\sigma_i\Theta_{i1}(m) \tag{5.24}$$

Remark 5.1 For the actual implementation of mixed time/event-triggered DMPC, if the event-triggering condition (5.24) is not satisfied during the prediction horizon N, k_i^d+N will be set as the next triggering instant and Problem 5.1 will be solved compulsively. That is, m is limited to $1\leqslant m\leqslant N-1$.

Remark 5.2 During the time instants $[k_i^d, k_i^d+m)$, the control input $\boldsymbol{u}_i^e(k_i^d+l)=\boldsymbol{u}_i^{e*}(k_i^d+l\,|\,k_i^d)(l=0,\dots,m-1)$ is applied to S_i^e. Hence, the uncertain item $\delta_i(m)$ is bounded by

$$\begin{aligned}\delta_i(m)=&\Big\|(\boldsymbol{A}_i^e)^m\boldsymbol{x}_i^e(k_i^d)+\sum_{l=1}^{m}(\boldsymbol{A}_i^e)^{m-l}\boldsymbol{B}_i^e\boldsymbol{u}_i^e(k_i^d+l-1)\\&+\sum_{l=1}^{m}(\boldsymbol{A}_i^e)^{m-l}\boldsymbol{w}_i^e(k_i^d+l-1)-(\boldsymbol{A}_i^e)^m\boldsymbol{x}_i^e(k_i^d)\\&-\sum_{l=1}^{m}(\boldsymbol{A}_i^e)^{m-l}\boldsymbol{B}_i^e\boldsymbol{u}_i^{e*}(k_i^d+l-1\,|\,k_i^d)\Big\|\\\leqslant&\frac{\|\boldsymbol{A}_i^e\|^m-1}{\|\boldsymbol{A}_i^e\|-1}\overline{\boldsymbol{w}}_i^e\end{aligned} \tag{5.25}$$

With the increasing of $\delta_i(m)$, the value of $\Theta_{i2}(m)\delta_i^2(m)+\Theta_{i3}(m)\delta_i(m)$ in (5.24) becomes larger and it is easier to meet the triggering condition. In addition, the upper bound of the prediction error as in (5.25) indicates the worst-case control performance.

Remark 5.3 Based on the robustness related constraints (5.9) and (5.15), we can observe from (5.21) and (5.22) that the influence of information received from time-triggered and event-triggered neighboring subsystems can be represented by different predicted state upper bounds.

Remark 5.4 By introducing $\delta_i(m)$, the designed event-triggering condition (5.24) only related to the local information of S_i^e itself. Thus, the event-triggering condition (5.24) can be easily determined without involving the information of its neighbors S_j^t and S_j^e.

5.3.2 Mixed Time/Event-Triggered Dual-Mode DMPC Algorithm

To further reduce the number of information exchanges, the dual-mode control scheme is adopted, i.e. if a subsystem enters its invariant sets, it is no demands for receiving predicted states of other subsystems and the state feedback control law is employed. The mixed time/event-triggered dual-mode DMPC algorithm is given as follows.

Algorithm 1 Mixed Time/Event-triggered Dual-mode DMPC

For subsystem S_i^e:

for $k=0,1,2,\ldots$ **do**

  if $k=0$ **then**

    if $\boldsymbol{x}_i^e(0)\in\phi_i^e$ **then**

      Sent $\boldsymbol{x}_i^e(0)$ to the controller and apply $\boldsymbol{u}_i^e(0)=\boldsymbol{K}_i^e\boldsymbol{x}_i^e(0)$. Inform subsystems S_j^t and S_j^e not to send information to subsystem S_i^e from this moment on. Transmit $(\boldsymbol{A}_i^e+\boldsymbol{B}_i^e\boldsymbol{K}_i^e)\boldsymbol{x}_i^e(0),\ldots,(\boldsymbol{A}_i^e+\boldsymbol{B}_i^e\boldsymbol{K}_i^e)^N\boldsymbol{x}_i^e(0)$ to subsystem S_j^t when $\boldsymbol{x}_j^t(0)\notin\phi_j^t$ and subsystem S_j^e when $\boldsymbol{x}_j^e(0)\notin\phi_j^e$.

    else

      Let $\hat{\boldsymbol{x}}_j^t(l\,|\,0)=0$ and $\hat{\boldsymbol{x}}_j^e(l\,|\,0)=0(l=0,\ldots,N-1)$, send $\boldsymbol{x}_i^e(0)$ to the controller. Solve Problem 5.1 based on $\boldsymbol{x}_i^e(0)$ to obtain the optimal control inputs $\boldsymbol{u}_i^{e*}(0\,|\,0),\ldots,\boldsymbol{u}_i^{e*}(N-1\,|\,0)$, and then apply $\boldsymbol{u}_i^{e*}(0\,|\,0)$. Transmit $\boldsymbol{x}_i^{e*}(1\,|\,0)$ and $\boldsymbol{u}_i^{e*}(1\,|\,0),\ldots,\boldsymbol{u}_i^{e*}(N-1\,|\,0)$ to S_j^t when $\boldsymbol{x}_j^t(0)\notin\phi_j^t$ and S_j^e when $\boldsymbol{x}_j^e(0)\notin\phi_j^e$.

    end if

  else

    if $\boldsymbol{x}_i^e(k)\in\phi_i^e$ **then**

      Send $\boldsymbol{x}_i^e(k)$ to the controller and apply $\boldsymbol{u}_i^e(k)=\boldsymbol{K}_i^e\boldsymbol{x}_i^e(k)$. Inform subsystems S_j^t and S_j^e not to send information to subsystem S_i^e from this moment on. Transmit $(\boldsymbol{A}_i^e+\boldsymbol{B}_i^e\boldsymbol{K}_i^e)\boldsymbol{x}_i^e(k),\ldots,(\boldsymbol{A}_i^e+\boldsymbol{B}_i^e\boldsymbol{K}_i^e)^N\boldsymbol{x}_i^e(k)$ to the neighbor S_j^t when $\boldsymbol{x}_j^t(k)\notin\phi_j^t$ and S_j^e when $\boldsymbol{x}_j^e(k)\notin\phi_j^e$.

    else

      if condition (5.24) is satisfied **then**

        Update $k_i^d=k$. Sent $\boldsymbol{x}_i^e(k_i^d)$ to the controller. Solve Problem 5.1 based on $\boldsymbol{x}_i^e(k_i^d)$ and the received information $\boldsymbol{x}_j^{t*}(k_i^d\,|\,k_i^d-1),\ldots,\boldsymbol{x}_j^{t*}(k_i^d+N-1\,|\,k_i^d-1)$ from subsystem S_j^t and

$\boldsymbol{x}_j^{e*}(k_i^d \mid k_j^{ei}),\dots,\boldsymbol{x}_j^{e*}(k_i^d+N-1 \mid k_j^{ei})$ from subsystem S_j^e to obtain the optimal control inputs $\boldsymbol{u}_i^{e*}(k_i^d \mid k_i^d),\dots,\boldsymbol{u}_i^{e*}(k_i^d+N-1 \mid k_i^d)$, and then apply $\boldsymbol{u}_i^{e*}(k_i^d \mid k_i^d)$. Transmit $\boldsymbol{x}_i^{e*}(k_i^d+1 \mid k_i^d)$ and $\boldsymbol{u}_i^{e*}(k_i^d+1 \mid k_i^d),\dots,\boldsymbol{u}_i^{e*}(k_i^d+N-1 \mid k_i^d)$ to the neighbor S_j^t when $\boldsymbol{x}_j^t(k_i^d)\notin\phi_j^t$ and S_j^e when $\boldsymbol{x}_j^e(k_i^d)\notin\phi_j^e$.
else
Apply $\bar{\boldsymbol{u}}_i^e(k \mid k)$ and do not transmit information.
end if
end if
end if
end for

For subsystem S_i^t:

for $k=0,1,2,\dots$ **do**
if $k=0$ **then**
if $\boldsymbol{x}_i^t(0)\in\phi_i^t$ **then**
Send $\boldsymbol{x}_i^t(0)$ to the controller and apply $\boldsymbol{u}_i^t(0)=\boldsymbol{K}_i^t\boldsymbol{x}_i^t(0)$. Inform subsystems S_j^t and S_j^e not to send information to subsystem S_i^t from this moment on. Transmit $(\boldsymbol{A}_i^t+\boldsymbol{B}_i^t\boldsymbol{K}_i^t)\boldsymbol{x}_i^t(0),\dots,(\boldsymbol{A}_i^t+\boldsymbol{B}_i^t\boldsymbol{K}_i^t)^N\boldsymbol{x}_i^t(0)$ to the neighbor S_j^t when $\boldsymbol{x}_j^t(0)\notin\phi_j^t$ and S_j^e when $\boldsymbol{x}_j^e(0)\notin\phi_j^e$.
else
Let $\hat{\boldsymbol{x}}_j^t(l \mid 0)=0$ and send $\hat{\boldsymbol{x}}_j^e(l \mid 0)=0(l=0,\dots,N-1),\boldsymbol{x}_i^t(0)$ to the controller. Solve Problem 5.2 based on $\boldsymbol{x}_i^t(0)$ to obtain the optimal control inputs $\boldsymbol{u}_i^{t*}(0 \mid 0),\dots,\boldsymbol{u}_i^{t*}(N-1 \mid 0)$, and then apply $\boldsymbol{u}_i^{t*}(0 \mid 0)$. Transmit $\boldsymbol{x}_i^{t*}(1 \mid 0)$ and $\boldsymbol{u}_i^{t*}(1 \mid 0),\dots,\boldsymbol{u}_i^{t*}(N-1 \mid 0)$ to the neighbor S_j^t when $\boldsymbol{x}_j^t(0)\notin\phi_j^t$ and S_j^e when $\boldsymbol{x}_j^e(0)\notin\phi_j^e$.
end if
else
if $\boldsymbol{x}_i^t(k)\in\phi_i^t$ **then**
Send $\boldsymbol{x}_i^t(k)$ to controller and apply $\boldsymbol{u}_i^t(k)=\boldsymbol{K}_i^t\boldsymbol{x}_i^t(k)$. Inform subsystems S_j^t and S_j^e not to send information to subsystem S_i^t from this moment on. Transmit $(\boldsymbol{A}_i^t+\boldsymbol{B}_i^t\boldsymbol{K}_i^t)\boldsymbol{x}_i^t(k),\dots,(\boldsymbol{A}_i^t+\boldsymbol{B}_i^t\boldsymbol{K}_i^t)^N\boldsymbol{x}_i^t(k)$ to the neighbor S_j^t when $\boldsymbol{x}_j^t(k)\notin\phi_j^t$ and S_j^e when $\boldsymbol{x}_j^e(k)\notin\phi_j^e$.
else
Send $\boldsymbol{x}_i^t(k)$ to the controller. Solve Problem 5.2 based on $\boldsymbol{x}_i^t(k)$ and the received information $\boldsymbol{x}_j^{t*}(k \mid k-1),\dots,\boldsymbol{x}_j^{t*}(k+N-1 \mid k-1)$ from subsystem S_j^t and $\boldsymbol{x}_j^{e*}(k \mid k_j^{ti}),\dots,\boldsymbol{x}_j^{e*}(k+N-1 \mid k_j^{ti})$ from subsystem S_j^e to obtain the optimal control inputs $\boldsymbol{u}_i^{t*}(k \mid k),\dots,\boldsymbol{u}_i^{t*}(k+N-1 \mid k)$, and then apply $\boldsymbol{u}_i^{t*}(k \mid k)$. Transmit $\boldsymbol{x}_i^{t*}(k+1 \mid k)$ and $\boldsymbol{u}_i^{t*}(k+1 \mid k),\dots,\boldsymbol{u}_i^{t*}(k+N-1 \mid k)$ to the neighbor S_j^t when $\boldsymbol{x}_j^t(k)\notin\phi_j^t$ and S_j^e when $\boldsymbol{x}_j^e(k)\notin\phi_j^e$.
end if
end if
end for

Remark 5.5 It is worth noting that the larger the triggering parameter σ_i, the stricter the event-triggering condition(5.24) will be, and the triggering frequency of problem computation will be reduced. In addition, in the framework of the dual-mode predictive control scheme, the larger the size of the invariant sets ϕ_i^e, the lower the computational consumption. Hence, the triggering parameter and the the size of the invariant sets should be well designed to achieve less resource usage in actual implementation.

5.4 Feasibility and Stability Analysis

In the proposed mixed time/event-triggered DMPC, different subsystems may have different problem computation frequency, thus the recursive feasibility of this DMPC optimization should be reconsidered. We analyze the recursive feasibility of Problems 5.1 and 5.2 in the following.

Theorem 5.1 *Problems 5.1 and 5.2 are recursively feasible if there exists disturbance upper bounds* $\overline{w}_i^e$ *and* $\overline{w}_i^t$, *and constants* α_i^e *and* α_i^t *satisfy the following inequalities:*

$$\overline{\boldsymbol{w}}_i^e \leqslant \frac{(\|\boldsymbol{A}_i^e\|-1)(1-\alpha_i^e)\varepsilon_i^e}{\sqrt{\overline{\lambda}(\boldsymbol{P}_i^e)}\,\|\boldsymbol{A}_i^e\|(\|\boldsymbol{A}_i^e\|^{N-1}-1)} \tag{5.26}$$

$$\overline{w}_i^t \leqslant \frac{(1-\alpha_i^t)\varepsilon_i^t}{\sqrt{\overline{\lambda}(\boldsymbol{P}_i^t)}\,\|\boldsymbol{A}_i^t\|^{N-1}} \tag{5.27}$$

$$\max\left\{\sqrt{1-\frac{\lambda(\overline{\boldsymbol{Q}}_i^e)}{\overline{\lambda}(\boldsymbol{P}_i^e)}},1-\frac{1}{1+\|\boldsymbol{A}_i^e\|^{N-2}(N-1)}\right\} \leqslant \alpha_i^e \tag{5.28}$$

$$\max\left\{\sqrt{1-\frac{\lambda(\overline{\boldsymbol{Q}}_i^t)}{\overline{\lambda}(\boldsymbol{P}_i^t)}},1-\frac{1}{N}\right\} \leqslant \alpha_i^t \tag{5.29}$$

Proof Suppose that Problem 5.1 is solved at time k_i^d, then we can obtain the optimal control inputs $\boldsymbol{u}_i^{e*}(k_i^d+l\,|\,k_i^d)$ and the predicted state $\boldsymbol{x}_i^{e*}(k_i^d+l\,|\,k_i^d)$. To verify the feasibility of Problem 5.1 at time k_i^d+m, one need to show the satisfaction of(5.8) and(5.9) with respect to control sequence(5.16)

(i) $\|\overline{\boldsymbol{x}}_i^e(k_i^d+l\,|\,k_i^d+m)\|_{\boldsymbol{P}_i^e}\leqslant\dfrac{N\alpha_i^e\varepsilon_i^e}{l-m}(l=m+1,\ldots,m+N)$.

First, we show that $\|\overline{\boldsymbol{x}}_i^e(k_i^d+m+N\,|\,k_i^d+m)\|_{\boldsymbol{P}_i^e}\leqslant\alpha_i^e\varepsilon_i^e$. From (5.18) and (5.25), we have

$$\|\overline{\boldsymbol{x}}_i^e(k_i^d+l\,|\,k_i^d+m)\|_{\boldsymbol{P}_i^e}-\|\boldsymbol{x}_i^{e*}(k_i^d+l\,|\,k_i^d)\|_{\boldsymbol{P}_i^e}\leqslant\frac{\sqrt{\overline{\lambda}(\boldsymbol{P}_i^e)}\,(\|\boldsymbol{A}_i^e\|^l-\|\boldsymbol{A}_i^e\|^{l-m})}{\|\boldsymbol{A}_i^e\|-1}\overline{\boldsymbol{w}}_i^e \tag{5.30}$$

Let $l=N$. According to the condition (5.26), we can further get $\|\overline{\boldsymbol{x}}_i^e(k_i^d+N\,|\,k_i^d+m)\|_{\boldsymbol{P}_i^e}\leqslant\varepsilon_i^e$, which implies $\overline{\boldsymbol{x}}_i^e(k_i^d+N\,|\,k_i^d+m)\in\phi_i^e$. Therefore, we have

$$\begin{aligned}
&\|\overline{\boldsymbol{x}}_i^e(k_i^d+N+1\,|\,k_i^d+m)\|_{\boldsymbol{P}_i^e}^2\\
&\leqslant\|\overline{\boldsymbol{x}}_i^e(k_i^d+N\,|\,k_i^d+m)\|_{\boldsymbol{P}_i^e}^2-\|\overline{\boldsymbol{x}}_i^e(k_i^d+N\,|\,k_i^d+m)\|_{\boldsymbol{Q}_i^e}^2,\\
&\qquad\vdots\\
&\|\overline{\boldsymbol{x}}_i^e(k_i^d+m+N\,|\,k_i^d+m)\|_{\boldsymbol{P}_i^e}^2\\
&\leqslant\|\overline{\boldsymbol{x}}_i^e(k_i^d+m+N-1\,|\,k_i^d+m)\|_{\boldsymbol{P}_i^e}^2-\\
&\|\overline{\boldsymbol{x}}_i^e(k_i^d+m+N-1\,|\,k_i^d+m)\|_{\boldsymbol{Q}_i^e}^2
\end{aligned}\tag{5.31}$$

Adding together all the expressions in (5.31) and adopting condition (5.28), we can then obtain that $\|\overline{\boldsymbol{x}}_i^e(k_i^d+m+N\,|\,k_i^d+m)\|_{\boldsymbol{P}_i^e}^2\leqslant\left[1-\dfrac{\underline{\lambda}(\boldsymbol{Q}_i^e)}{\overline{\lambda}(\boldsymbol{P}_i^e)}\right](\varepsilon_i^e)^2\leqslant(\alpha_i^e\varepsilon_i^e)^2$. Subsequently, we show that $\|\overline{\boldsymbol{x}}_i^e(k_i^d+l\,|\,k_i^d+m)\|_{\boldsymbol{P}_i^e}\leqslant\dfrac{N\alpha_i^e\varepsilon_i^e}{l-m}(l=m+1,\ldots,m+N-1)$. As can be seen from (5.30), the above expression is equivalent to proving $\dfrac{\sqrt{\overline{\lambda}(\boldsymbol{P}_i^e)}\,(\|\boldsymbol{A}_i^e\|^l-\|\boldsymbol{A}_i^e\|^{l-m})}{\|\boldsymbol{A}_i^e\|-1}\overline{\boldsymbol{w}}_i^e\leqslant\dfrac{mN\alpha_i^e\varepsilon_i^e}{l(l-m)}$. Applying the condition (5.26), we can get $\dfrac{\sqrt{\overline{\lambda}(\boldsymbol{P}_i^e)}\,(\|\boldsymbol{A}_i^e\|^l-\|\boldsymbol{A}_i^e\|^{l-m})}{\|\boldsymbol{A}_i^e\|-1}\overline{\boldsymbol{w}}_i^e\leqslant\|\boldsymbol{A}_i^e\|^{N-2}(1-\alpha_i^e)\varepsilon_i^e$. Noting $l\leqslant m+N-1$ and $m\geqslant1$, it is easy to show that $\dfrac{mN\alpha_i^e\varepsilon_i^e}{l(l-m)}\geqslant\dfrac{\alpha_i^e\varepsilon_i^e}{N-1}$. Furthermore, it follows from (5.28) that $\|\boldsymbol{A}_i^e\|^{N-2}(1-\alpha_i^e)\varepsilon_i^e\leqslant\dfrac{\alpha_i^e\varepsilon_i^e}{N-1}$.

(ii) $\overline{\boldsymbol{u}}_i^e(k_i^d+l|k_i^d+m)\in\boldsymbol{U}_i^e(l=m,\ldots,m+N-1)$.

From(5.16), we have $\overline{\boldsymbol{u}}_i^e(k_i^d+l|k_i^d+m)=\boldsymbol{u}_i^{e*}(k_i^d+l|k_i^d)\in\boldsymbol{U}_i^e(l=m,\ldots,N-1)$. Since $\overline{\boldsymbol{x}}_i^e(k_i^d+l|k_i^d+m)\in\phi_i^e(l=N,\ldots,m+N-1)$, we obtain $\overline{\boldsymbol{u}}_i^e(k_i^d+l|k_i^d+m)=\boldsymbol{K}_i^e\overline{\boldsymbol{x}}_i^e(k_i^d+l|k_i^d+m)\in\boldsymbol{U}_i^e(l=N,\ldots,m+N-1)$.

The recursive feasibility of Problem 5.2 is proven similarly in the sequel. Assume Problem 5.2 is solved at time k, then the optimal control sequence $\boldsymbol{u}_i^{t*}(k+l|k)$ and the predicted state $\boldsymbol{x}_i^{t*}(k+l|k)$ can be obtained. Consider the control sequence $\overline{\boldsymbol{u}}_i^t(k+l|k+1)$ at time $k+1$:

$$\overline{\boldsymbol{u}}_i^t(k+l|k+1)=\begin{cases}\boldsymbol{u}_i^{t*}(k+l|k),l=1,\ldots,N-1\\ \boldsymbol{K}_i^t\overline{\boldsymbol{x}}_i^t(k+l|k+1),l=N\end{cases}\tag{5.32}$$

then Problem 5.2 is feasible at time $k+1$ if(5.32)satisfies the constraints (5.14)and(5.15).

(iii) $\|\overline{\boldsymbol{x}}_i^t(k+l|k+1)\|_{\boldsymbol{P}_i^t}\leqslant\dfrac{N\alpha_i^t\varepsilon_i^t}{l-1}(l=2,\ldots,N+1)$.

First, we show that $\|\overline{\boldsymbol{x}}_i^t(k+N+1|k+1)\|_{\boldsymbol{P}_i^t}\leqslant\alpha_i^t\varepsilon_i^t$. Using the condition (5.27), we have $\|\overline{\boldsymbol{x}}_i^t(k+N|k+1)\|_{\boldsymbol{P}_i^t}\leqslant\|\overline{\boldsymbol{x}}_i^{t*}(k+N|k)\|_{\boldsymbol{P}_i^t}+\sqrt{\overline{\lambda}(\boldsymbol{P}_i^t)}\|\boldsymbol{A}_i^t\|^{N-1}\overline{\boldsymbol{w}}_i^t\leqslant\varepsilon_i^t$. Furthermore, it follows from (5.29) that $\|\overline{\boldsymbol{x}}_i^t(k+N+1|k+1)\|_{\boldsymbol{P}_i^t}^2\leqslant\left[1-\dfrac{\lambda(\overline{\boldsymbol{Q}}_i^t)}{\overline{\lambda}(\boldsymbol{P}_i^t)}\right](\varepsilon_i^t)^2\leqslant(\alpha_i^t\varepsilon_i^t)^2$. Next, we show that $\|\overline{\boldsymbol{x}}_i^t(k+l|k+1)\|_{\boldsymbol{P}_i^t}\leqslant\dfrac{N\alpha_i^t\varepsilon_i^t}{l-1}(l=2,\ldots,N)$, which is equivalent to proving $\sqrt{\overline{\lambda}(\boldsymbol{P}_i^t)}\|\boldsymbol{A}_i^t\|^{l-1}\overline{\boldsymbol{w}}_i^t\leqslant\dfrac{N\alpha_i^t\varepsilon_i^t}{l(l-1)}$. According to conditions(5.27)and(5.29), we obtain $\sqrt{\overline{\lambda}(\boldsymbol{P}_i^t)}\|\boldsymbol{A}_i^t\|^{l-1}\overline{\boldsymbol{w}}_i^t\leqslant(1-\alpha_i^t)\varepsilon_i^t\leqslant\dfrac{\alpha_i^t\varepsilon_i^t}{N-1}\leqslant\dfrac{N\alpha_i^t\varepsilon_i^t}{l(l-1)}$.

(iv) $\overline{\boldsymbol{u}}_i^t(k+l|k+1)\in\boldsymbol{U}_i^t(l=1,\ldots,N)$.

From(5.32), we have $\overline{\boldsymbol{u}}_i^t(k+l|k+1)=\boldsymbol{u}_i^{t*}(k+l|k)\in\boldsymbol{U}_i^t(l=1,\ldots,N-1)$. Since $\overline{\boldsymbol{x}}_i^t(k+N|k+1)\in\phi_i^t$, we can get $\overline{\boldsymbol{u}}_i^t(k+N|k+1)=\boldsymbol{K}_i^t\overline{\boldsymbol{x}}_i^t(k+N|k+1)\in\boldsymbol{U}_i^t$.

□

The stability of the system under the proposed mixed triggered DMPC is analyzed in the following.

Theorem 5.2 *Subsystems S_i^e and S_i^t will steer to the invariant sets ϕ_i^e and ϕ_i^t, and gradually get into the disturbance invariant sets $\psi_i^e=\{x_i^e\in\mathbb{R}^{n_i}:\|x_i^e\|_{P_i^e}^2\leqslant(\eta_i^e\varepsilon_i^e)^2\}$ and $\psi_i^t=\{x_i^t\in\mathbb{R}^{n_i}:\|x_i^t\|_{P_i^t}^2\leqslant(\eta_i^t\varepsilon_i^t)^2\}$ in finite time, respectively, if there exist disturbance upper bounds $\overline{w}_i^e$ and $\overline{w}_i^t$, cooperation matrices $Q_{ij}^{ee}>0$, $Q_{ij}^{et}>0$, $Q_{ij}^{te}>0$ and $Q_{ij}^{tt}>0$, constants α_i^e, α_i^t, $0<\eta_i^e<1$, $0<\eta_i^t<1$, $\mu_i^e>0$ and $\mu_i^t>0$ such that the following conditions hold:*

$$\frac{\underline{\lambda}(Q_i^e)}{\overline{\lambda}(P_i^e)}(\varepsilon_i^e)^2-\overline{\Gamma}_{i3}^e-\overline{\Gamma}_{i4}^e-\overline{\Gamma}_{i5}^e\geqslant 0 \tag{5.33}$$

$$-\overline{\Theta}_{i1}+\overline{\Theta}_{i2}(\overline{w}_i^t)^2+\overline{\Theta}_{i3}\overline{w}_i^t\leqslant 0 \tag{5.34}$$

$$\overline{w}_i^e\leqslant\min\left\{\frac{(\|A_i^e\|-1)(1-\alpha_i^e)\varepsilon_i^e}{\sqrt{\overline{\lambda}(P_i^e)}\|A_i^e\|(\|A_i^e\|^{N-1}-1)},\sqrt{\frac{\Pi_i^e}{\Xi_i^e}}\eta_i^e\varepsilon_i^e\right\} \tag{5.35}$$

$$\overline{w}_i^t\leqslant\min\left\{\frac{(1-\alpha_i^t)\varepsilon_i^t}{\sqrt{\overline{\lambda}(P_i^t)}\|A_i^t\|^{N-1}},\sqrt{\frac{\Pi_i^t}{\Xi_i^t}}\eta_i^t\varepsilon_i^t\right\} \tag{5.36}$$

$$\frac{[\underline{\lambda}(\overline{Q}_i^e)-\overline{\lambda}(P_i^e)]\underline{\lambda}(P_i^e)}{\overline{\lambda}(P_i^e)}<\mu_i^e<\frac{\underline{\lambda}(\overline{Q}_i^e)\underline{\lambda}(P_i^e)}{\overline{\lambda}(P_i^e)} \tag{5.37}$$

$$\frac{[\underline{\lambda}(\overline{Q}_i^t)-\overline{\lambda}(P_i^t)]\underline{\lambda}(P_i^t)}{\overline{\lambda}(P_i^t)}<\mu_i^t<\frac{\underline{\lambda}(\overline{Q}_i^t)\underline{\lambda}(P_i^t)}{\overline{\lambda}(P_i^t)} \tag{5.38}$$

where

$$\begin{aligned}\overline{\Gamma}_{i3}^e=&\sum_{j\in\mathcal{N}_i^e}\left[\sqrt{\frac{\overline{\lambda}(Q_{ij}^{et})}{\underline{\lambda}(P_i^e)}}(N\alpha_i^e+1-\alpha_i^e)\varepsilon_i^e+\frac{\sqrt{\overline{\lambda}(Q_{ij}^{et})}N\alpha_j^t\varepsilon_j^t}{\sqrt{\underline{\lambda}(P_j^t)}}\right]^2\\&+\sum_{l=1}^{N-1}\sum_{j\in\mathcal{N}_i^e}\left[\frac{\sqrt{\overline{\lambda}(Q_{ij}^{et})}N\alpha_i^e\varepsilon_i^e}{\sqrt{\underline{\lambda}(P_i^e)}l}+\frac{\sqrt{\overline{\lambda}(Q_{ij}^{et})}N\alpha_j^t\varepsilon_j^t}{\sqrt{\underline{\lambda}(P_j^t)}(l+1)}\right]^2,\\\overline{\Gamma}_{i4}^e=&\sum_{j\in\mathcal{N}_i^e}\left[\sqrt{\frac{\overline{\lambda}(Q_{ij}^{ee})}{\underline{\lambda}(P_i^e)}}(N\alpha_i^e+1-\alpha_i^e)\varepsilon_i^e+\frac{\sqrt{\overline{\lambda}(Q_{ij}^{ee})}N\alpha_j^e\varepsilon_j^e}{\sqrt{\underline{\lambda}(P_j^e)}}\right]^2\\&+\sum_{l=1}^{N-1}\sum_{j\in\mathcal{N}_i^e}\left[\frac{\sqrt{\overline{\lambda}(Q_{ij}^{ee})}N\alpha_i^e\varepsilon_i^e}{\sqrt{\underline{\lambda}(P_i^e)}l}+\frac{\sqrt{\overline{\lambda}(Q_{ij}^{ee})}N\alpha_j^e\varepsilon_j^e}{\sqrt{\underline{\lambda}(P_j^e)}(l+1)}\right]^2,\end{aligned}$$

$$\overline{\Gamma}_{i5}^{e}=\left[\frac{\overline{\boldsymbol{w}}_i^e(\|\boldsymbol{A}_i^e\|^{N-2}-1)}{\|\boldsymbol{A}_i^e\|-1}\right]^2\cdot\left[\overline{\lambda}(\boldsymbol{Q}_i^e)\sum_{l=2}^{N-1}\|\boldsymbol{A}_i^e\|^{2(l-1)}+\overline{\lambda}(\boldsymbol{P}_i^e)\|\boldsymbol{A}_i^e\|^{2(N-1)}\right]$$

$$+\frac{2N\alpha_i^e\varepsilon_i^e\overline{\boldsymbol{w}}_i^e(\|\boldsymbol{A}_i^e\|^{N-2}-1)}{\|\boldsymbol{A}_i^e\|-1}\left[\frac{\overline{\lambda}(\boldsymbol{Q}_i^e)}{\sqrt{\underline{\lambda}(\boldsymbol{P}_i^e)}}\sum_{l=2}^{N-1}\frac{\|\boldsymbol{A}_i^e\|^{l-1}}{l-1}+\frac{\sqrt{\overline{\lambda}(\boldsymbol{P}_i^e)}\|\boldsymbol{A}_i^e\|^{N-1}}{2}\right],$$

$$\overline{\Theta}_{i1}=\frac{\underline{\lambda}(\boldsymbol{Q}_i^t)}{\overline{\lambda}(\boldsymbol{P}_i^t)}(\varepsilon_i^t)^2-\sum_{j\in\mathcal{N}_i^t}\left[\sqrt{\frac{\overline{\lambda}(\boldsymbol{Q}_{ij}^{tt})}{\underline{\lambda}(\boldsymbol{P}_i^t)}}(N\alpha_i^t+1-\alpha_i^t)\varepsilon_i^t+\frac{\sqrt{\overline{\lambda}(\boldsymbol{Q}_{ij}^{tt})}N\alpha_j^t\varepsilon_j^t}{\sqrt{\underline{\lambda}(\boldsymbol{P}_j^t)}}\right]^2$$

$$-\sum_{l=1}^{N-1}\sum_{j\in\mathcal{N}_i^t}\left[\frac{\sqrt{\overline{\lambda}(\boldsymbol{Q}_{ij}^{tt})}N\alpha_i^t\varepsilon_i^t}{\sqrt{\underline{\lambda}(\boldsymbol{P}_i^t)}\,l}+\frac{\sqrt{\overline{\lambda}(\boldsymbol{Q}_{ij}^{tt})}N\alpha_j^t\varepsilon_j^t}{\sqrt{\underline{\lambda}(\boldsymbol{P}_j^t)}(l+1)}\right]^2$$

$$-\sum_{j\in\mathcal{N}_i^t}\left[\sqrt{\frac{\overline{\lambda}(\boldsymbol{Q}_{ij}^{te})}{\underline{\lambda}(\boldsymbol{P}_i^t)}}(N\alpha_i^t+1-\alpha_i^t)\varepsilon_i^t+\frac{\sqrt{\overline{\lambda}(\boldsymbol{Q}_{ij}^{te})}N\alpha_j^e\varepsilon_j^e}{\sqrt{\underline{\lambda}(\boldsymbol{P}_j^e)}}\right]^2$$

$$-\sum_{l=1}^{N-1}\sum_{j\in\mathcal{N}_i^t}\left[\frac{\sqrt{\overline{\lambda}(\boldsymbol{Q}_{ij}^{te})}N\alpha_i^t\varepsilon_i^t}{\sqrt{\underline{\lambda}(\boldsymbol{P}_i^t)}\,l}+\frac{\sqrt{\overline{\lambda}(\boldsymbol{Q}_{ij}^{te})}N\alpha_j^e\varepsilon_j^e}{\sqrt{\underline{\lambda}(\boldsymbol{P}_j^e)}(l+1)}\right]^2,$$

$$\overline{\Theta}_{i2}=\overline{\lambda}(\boldsymbol{Q}_i^t)\sum_{l=1}^{N-1}\|\boldsymbol{A}_i^t\|^{2(l-1)}+\overline{\lambda}(\boldsymbol{P}_i^t)\|\boldsymbol{A}_i^t\|^{2(N-1)},$$

$$\overline{\Theta}_{i3}=\frac{2\overline{\lambda}(\boldsymbol{Q}_i^t)N\alpha_i^t\varepsilon_i^t}{\sqrt{\underline{\lambda}(\boldsymbol{P}_i^t)}}\sum_{l=1}^{N-1}\frac{\|\boldsymbol{A}_i^t\|^{l-1}}{l}+2\sqrt{\overline{\lambda}(\boldsymbol{P}_i^t)}\alpha_i^t\varepsilon_i^t\|\boldsymbol{A}_i^t\|^{N-1},$$

$$\Pi_i^e=\frac{\underline{\lambda}(\overline{\boldsymbol{Q}}_i^e)}{\overline{\lambda}(\boldsymbol{P}_i^e)}-\frac{\boldsymbol{\mu}_i^e}{\underline{\lambda}(\boldsymbol{P}_i^e)},\Xi_i^e=\overline{\lambda}(\boldsymbol{P}_i^e)+\frac{\|(\boldsymbol{A}_i^e+\boldsymbol{B}_i^e\boldsymbol{K}_i^e)^T\boldsymbol{P}_i^e\|^2}{\boldsymbol{\mu}_i^e},$$

$$\Pi_i^t=\frac{\underline{\lambda}(\overline{\boldsymbol{Q}}_i^t)}{\overline{\lambda}(\boldsymbol{P}_i^t)}-\frac{\boldsymbol{\mu}_i^t}{\underline{\lambda}(\boldsymbol{P}_i^t)},\Xi_i^t=\overline{\lambda}(\boldsymbol{P}_i^t)+\frac{\|(\boldsymbol{A}_i^t+\boldsymbol{B}_i^t\boldsymbol{K}_i^t)^T\boldsymbol{P}_i^t\|^2}{\boldsymbol{\mu}_i^t}.$$

Proof The convergence of S_i^e indicates that the cost function $J_i^e(k)$ decreases monotonically. Therefore, $\Theta_{i1}(m)$ should satisfy $\Theta_{i1}(m)\geqslant 0$, which will be verified in the following. Since $\boldsymbol{x}_i^e(k_i^d+m-1)\notin\phi_i^e$, we have

$$\|\boldsymbol{x}_i^e(k_i^d+m-1)\|_{\boldsymbol{Q}_i^e}^2\geqslant\frac{\underline{\lambda}(\boldsymbol{Q}_i^e)}{\overline{\lambda}(\boldsymbol{P}_i^e)}\|\boldsymbol{x}_i^e(k_i^d+m-1)\|_{\boldsymbol{P}_i^e}^2\geqslant\frac{\underline{\lambda}(\boldsymbol{Q}_i^e)}{\overline{\lambda}(\boldsymbol{P}_i^e)}(\varepsilon_i^e)^2 \quad (5.39)$$

By means of (5.30) and let $l=m$, we can get the inequality in the following:

$$\|\boldsymbol{x}_i^e(k_i^d+m)\|_{\boldsymbol{P}_i^e}\leqslant\frac{N\alpha_i^e\varepsilon_i^e}{m}+\frac{\sqrt{\overline{\lambda}(\boldsymbol{P}_i^e)}(\|\boldsymbol{A}_i^e\|^m-1)}{\|\boldsymbol{A}_i^e\|-1}\overline{\boldsymbol{w}}_i^e\leqslant(N\alpha_i^e+1-\alpha_i^e)\varepsilon_i^e \quad (5.40)$$

Based on (5.40) and $k_i^d+l-k_j^{ei+m}\geqslant l-m+1$, it follows that $\overline{\Gamma}_{i3}^e(m)\leqslant\overline{\Gamma}_{i3}^e$ and $\overline{\Gamma}_{i4}^e(m)\leqslant\overline{\Gamma}_{i4}^e$. Moreover, we have

$$\overline{\Gamma}_{i5}^e(m)\leqslant\sum_{l=m}^{N-1}[\|\boldsymbol{x}_i^{e*}(k_i^d+l\,|\,k_i^d)\|_{\boldsymbol{Q}_i^e}^2-\|\overline{\boldsymbol{x}}_i^e(k_i^d+l\,|\,k_i^d+m-1)\|_{\boldsymbol{Q}_i^e}^2]$$
$$+\|\boldsymbol{x}_i^{e*}(k_i^d+N\,|\,k_i^d)\|_{\boldsymbol{P}_i^e}^2-\|\overline{\boldsymbol{x}}_i^e(k_i^d+N\,|\,k_i^d+m-1)\|_{\boldsymbol{P}_i^e}^2\leqslant\overline{\Gamma}_{i5}^e \tag{5.41}$$

From the above and based on (5.33), we obtain $\Theta_{i1}(m)\geqslant\dfrac{\underline{\lambda}(\boldsymbol{Q}_i^e)}{\overline{\lambda}(\boldsymbol{P}_i^e)}(\varepsilon_i^e)^2-\overline{\Gamma}_{i3}^e-\overline{\Gamma}_{i4}^e-\overline{\Gamma}_{i5}^e\geqslant0$. Thus, if the triggering condition (5.24) is not satisfied, we have $\overline{J}_i^e(k_i^d+m)-\overline{J}_i^e(k_i^d+m-1)<0$. If the triggering condition (5.24) is satisfied, update $k_i^d=k_i^d+m$ and reset the error $\delta_i(m)=\|\boldsymbol{x}_i^e(k_i^d+m)-\boldsymbol{x}_i^{e*}(k_i^d+m\,|\,k_i^d+m)\|$ to 0, and we have $J_i^{e*}(k_i^d+m)-\overline{J}_i^e(k_i^d+m-1)\leqslant-\Theta_{i1}(m)\leqslant0$. Based on the above discussion, we can conclude that subsystem S_i^e will steer to the invariant set ϕ_i^e.

We then analyze the convergence of S_i^t in the sequel. Let $J_i^{t*}(k)$ represent the cost function obtained based on the optimal control inputs at time k and let $\overline{J}_i^t(k+1)$ indicate the cost function based on control inputs (5.32). Similarly to the analysis in the event-triggered case, we obtain $\overline{J}_i^t(k+1)-J_i^{t*}(k)\leqslant-\overline{\Theta}_{i1}+\overline{\Theta}_{i2}(\overline{\boldsymbol{w}}_i^t)^2+\overline{\Theta}_{i3}\,\overline{\boldsymbol{w}}_i^t$. Employing condition (5.34), we have $\overline{J}_i^t(k+1)\leqslant J_i^{t*}(k)$. Therefore, subsystem S_i^t will converge to the invariant set ϕ_i^t.

Next, we prove the closed-loop stability of S_i^e controlled by the state feedback control law $\boldsymbol{K}_i^e$. Assume that $\boldsymbol{x}_i^e(\widetilde{k}_i)\in\phi_i^e$, so we have $\boldsymbol{u}_i^e(k)=\boldsymbol{K}_i^e\boldsymbol{x}_i^e(k)$ $(k\geqslant\widetilde{k}_i)$. Choose the Lyapunov function as $\boldsymbol{V}_i^e(k)=\|\boldsymbol{x}_i^e(k)\|_{\boldsymbol{P}_i^e}^2$, then the difference between $\boldsymbol{V}_i^e(k+1)$ and $\boldsymbol{V}_i^e(k)$ is given by

$$\begin{aligned}\Delta\boldsymbol{V}_i^e(k)&=\boldsymbol{V}_i^e(k+1)-\boldsymbol{V}_i^e(k)\\&\leqslant-\|\boldsymbol{x}_i^e(k)\|_{\overline{\boldsymbol{Q}}_i^e}^2+\|\boldsymbol{w}_i^e(k)\|_{\boldsymbol{P}_i^e}^2+2[\boldsymbol{x}_i^e(k)]^{\mathrm{T}}(\boldsymbol{A}_i^e+\boldsymbol{B}_i^e\boldsymbol{K}_i^e)^{\mathrm{T}}\boldsymbol{P}_i^e\boldsymbol{w}_i^e(k)\\&\leqslant-\Pi_i^e\|\boldsymbol{x}_i^e(k)\|_{\boldsymbol{P}_i^e}^2+\Xi_i^e(\overline{\boldsymbol{w}}_i^e)^2\end{aligned} \tag{5.42}$$

Considering the condition (5.35), we can further get

$$\Delta V_i^e(k) \leqslant \Pi_i^e\left[-\|x_i^e(k)\|_{P_i^e}^2 + (\eta_i^e \varepsilon_i^e)^2\right] \tag{5.43}$$

Assume that there is no time instant $k \geqslant \tilde{k}_i$ such that $x_i^e(k) \in \psi_i^e$. Thus, $\forall k \geqslant \tilde{k}_i$, there have to be a constant $\epsilon_i^e > 0$ such that

$$\|x_i^e(k)\|_{P_i^e}^2 \geqslant (\eta_i^e \varepsilon_i^e)^2 + \epsilon_i^e \tag{5.44}$$

Combining(5.43)and(5.44), we obtain

$$\begin{aligned} V_i^e(\tilde{k}_i+1) - V_i^e(\tilde{k}_i) &\leqslant -\Pi_i^e \epsilon_i^e, \\ &\vdots \\ V_i^e(k) - V_i^e(k-1) &\leqslant -\Pi_i^e \epsilon_i^e \end{aligned} \tag{5.45}$$

Adding together all the inequalities in(5.45), it follows that $V_i^e(k) \leqslant (\varepsilon_i^e)^2 - (k - \tilde{k}_i)\Pi_i^e \epsilon_i^e$. Let $\bar{k}_i = \inf\left\{k \in \mathbb{N} : k \geqslant \tilde{k}_i + \frac{(\varepsilon_i^e)^2 - (\eta_i^e \varepsilon_i^e)^2}{\Pi_i^e \epsilon_i^e}\right\}$, then we have $\|x_i^e(\bar{k}_i)\|_{P_i^e}^2 \leqslant (\eta_i^e \varepsilon_i^e)^2$, which contradicts with(5.44). Therefore, there is a time instant $\bar{k}_i \geqslant \bar{k}_i$ such that $x_i^e(\bar{k}_i) \in \psi_i^e$. In addition, we can observe from(5.37) that $0 < \Pi_i^e < 1$. Hence, we have

$$\|x_i^e(\bar{k}_i+1)\|_{P_i^e}^2 \leqslant \|x_i^e(\bar{k}_i)\|_{P_i^e}^2 + \Pi_i^e\left[-\|x_i^e(\bar{k}_i)\|_{P_i^e}^2 + (\eta_i^e \varepsilon_i^e)^2\right] \leqslant (\eta_i^e \varepsilon_i^e)^2 \tag{5.46}$$

We can then obtain that ψ_i^e is a disturbance invariant set for S_i^e. Therefore, the subsystem S_i^e will converge into the disturbance invariant set ψ_i^e in finite time. Based on conditions(5.36)and(5.38)and similar to the proof above, one can see that subsystem S_i^t will get into its disturbance invariant set ψ_i^t in finite time. □

5.5 Example

In this section, a multi-agent control system is considered [10], where agent 1 and agent 2 are event-triggered while agent 3 is time-triggered. The dynamics of each agent i $(i=1,2,3)$ can be described as:

$$x_i(k+1) = \begin{bmatrix} 1 & 0.1 & 0 & 0 \\ 0 & 1 & 0 & 0 \\ 0 & 0 & 1 & 0.1 \\ 0 & 0 & 0 & 1 \end{bmatrix} x_i(k) + \begin{bmatrix} 0.05 & 0 \\ 1 & 0 \\ 0 & 0.05 \\ 0 & 1 \end{bmatrix} u_i(k) + w_i(k) \tag{5.47}$$

where $\boldsymbol{x}_i = [x_{i1}, x_{i2}, x_{i3}, x_{i4}]^{\mathrm{T}}$, x_{i1} and x_{i3} denote the horizonal position s_i^x and vertical position s_i^y of agent i, respectively; x_{i2} is the horizonal velocity v_i^x and x_{i4} is the vertical velocity v_i^y; $\boldsymbol{u}_i = [u_{i1}, u_{i2}]^{\mathrm{T}}$, u_{i1} and u_{i2} represent the horizonal force u_i^x and vertical force u_i^y, respectively.

The control input is bounded as $\|\boldsymbol{u}_i(k)\|_\infty \leqslant 0.12$ and we design the parameters in Problems 5.1 and 5.2 as $N=10$, $\boldsymbol{Q}_i = \boldsymbol{I}_{4\times 4}$, $\boldsymbol{R}_i = 5\boldsymbol{I}_{2\times 2}$, $\varepsilon_1 = \varepsilon_2 = 1$, $\varepsilon_3 = 1.2$, $\mathcal{N}_1 = \{2,3\}$, $\mathcal{N}_2 = \{1,3\}$, $\mathcal{N}_3 = \{1,2\}$. The following terminal weighting matrix $\boldsymbol{P}_i$ and the feedback control law $\boldsymbol{K}_i$ are obtained based on Riccatti equations:

$$\boldsymbol{P}_i = \begin{bmatrix} 12.5862 & 2.2913 & 0 & 0 \\ 2.2913 & 3.2693 & 0 & 0 \\ 0 & 0 & 12.5862 & 2.2913 \\ 0 & 0 & 2.2913 & 3.2693 \end{bmatrix},$$

$$\boldsymbol{K}_i = \begin{bmatrix} -0.3424 & -0.4309 & 0 & 0 \\ 0 & 0 & -0.3424 & -0.4309 \end{bmatrix}.$$

The upper bounds of disturbances are given as $\overline{w}_1 = \overline{w}_2 = 2\times 10^{-5}$ and $\overline{w}_3 = 7\times 10^{-4}$ based on Theorems 5.1 and 5.2; the parameter α_i can be determined as 0.9611 and the cooperation matrix is $\boldsymbol{Q}_{ij} = [10^{-4}, 0, 0, 0; 0, 0, 0, 0; 0, 0, 10^{-4}, 0; 0, 0, 0, 0]$ $(j \in \mathcal{N}_i)$. The triggering parameters for agent 1 and agent 2 are $\sigma_1 = 5\times 10^{-3}$ and $\sigma_2 = 3\times 10^{-3}$, respectively. The initial conditions are given by $\boldsymbol{x}_1(0) = [0.4, 0, -0.5, 0]^{\mathrm{T}}$, $\boldsymbol{x}_2(0) = [0.3, 0, 0.6, 0]^{\mathrm{T}}$, $\boldsymbol{x}_3(0) = [-0.5, 0, -0.4, 0]^{\mathrm{T}}$.

Fig. 5.2 illustrates the trajectories of three agents under the proposed mixed triggered DMPC, where each agent converges to the equilibrium point in a finite time. The velocities and forces of each agent are shown in Figs. 5.3 and 5.4, respectively. Fig. 5.5 shows the triggering instants of three agents, where value 1 indicates that the triggering condition is satisfied and value 0 indicates that the triggering condition is not satisfied. Before entering the invariant sets, agent 1 and agent 2 are triggered aperiodically, which reduces the communication and computation burden, and agent 3 is triggered periodically. After agents enter their invariant sets, there is no triggering instant. We further compare the control performance of agents under three different kinds of control schemes in Fig. 5.6, in which the performance index is defined by $J(t) \triangleq \sum_{i=1}^{M} \sum_{k=0}^{t} \left[\| \boldsymbol{x}_i(k) \|_{\boldsymbol{Q}_i}^2 + \| \boldsymbol{u}_i(k) \|_{\boldsymbol{R}_i}^2 + \sum_{j \in \mathcal{N}_i} \| \boldsymbol{x}_i(k) - \boldsymbol{x}_j(k) \|_{\boldsymbol{Q}_{ij}}^2 \right]$. As shown in Fig. 5.6, the control performance under mixed time/event-triggered DMPC is better than that under the event-triggered DMPC and worse than in the time-triggered case. Finally, we illustrate the problem-solving frequency of agents under three different kinds of control schemes in Table 5.1, which indicates that the mixed time/event-triggered DMPC saves more resources than TT-DMPC.

Table 5.1 Problem-solving frequency of DMPC optimization problem under three different triggered mechanisms

Triggered scheme	Agent 1	Agent 2	Agent 3
Time-triggered DMPC	10	10	10
Mixed triggered DMPC	2	5	9
Event-triggered DMPC	2	4	3

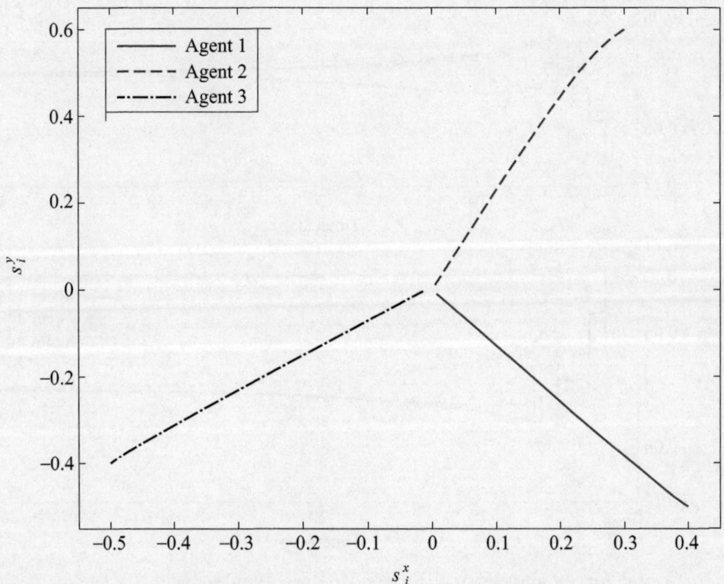

Fig. 5.2 Trajectories of three agents

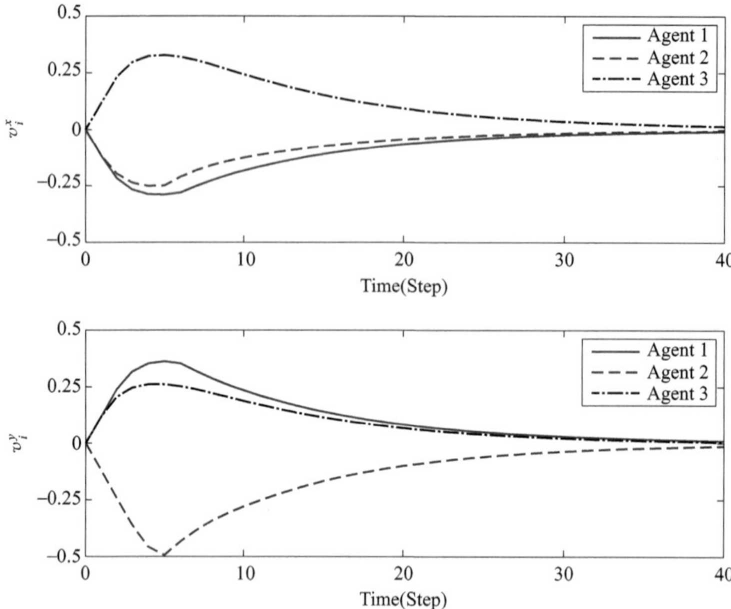

Fig. 5. 3 Horizontal velocities and vertical velocities of three agents

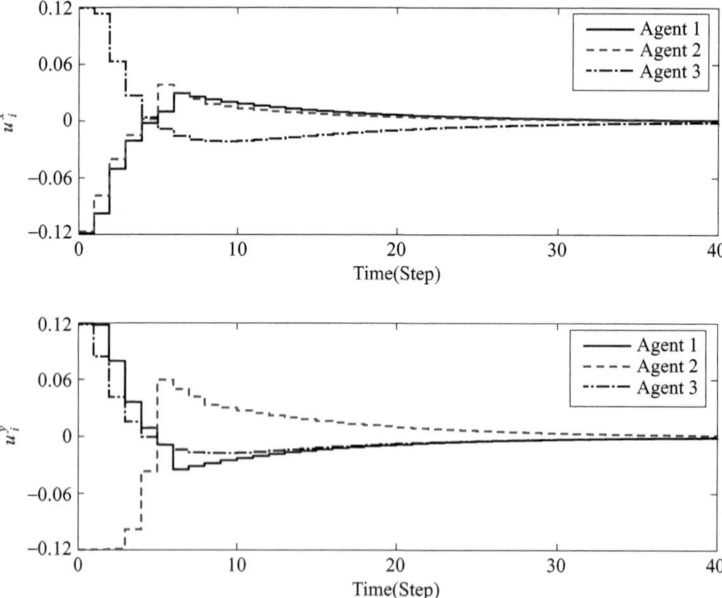

Fig. 5. 4 Horizontal forces and vertical forces of three agents

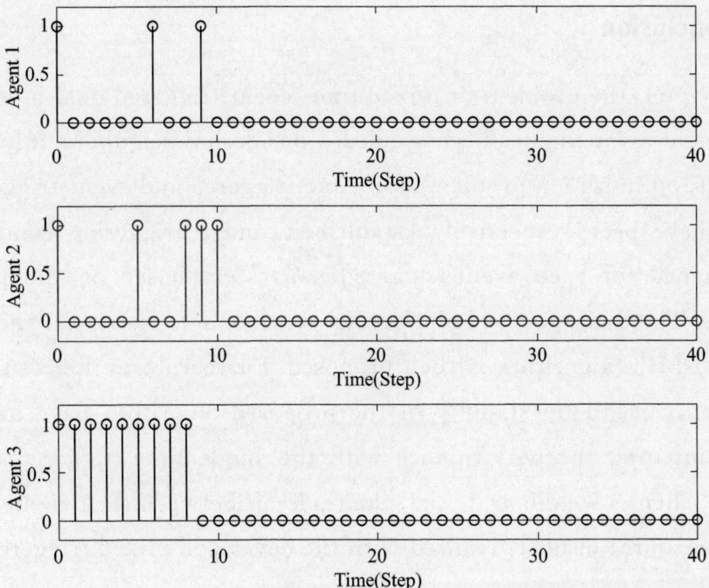

Fig. 5. 5 Triggering instants for three agents

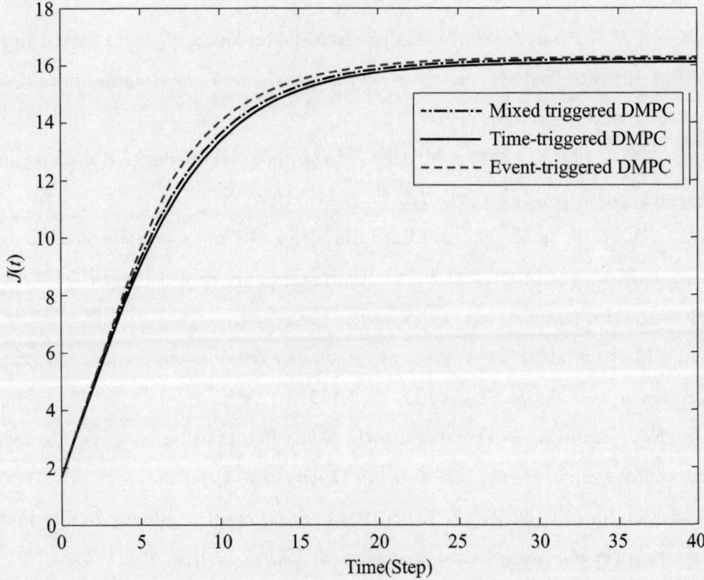

Fig. 5. 6 Overall control performance indices under three different triggered mechanisms

5.6 Conclusion

In this chapter, the problem of mixed time/event-triggered dual-mode DMPC for large-scale systems has been studied. Considering neighbors' information, the DMPC optimization problems for time-triggered and event-triggered subsystems have been respectively established, and a triggering condition has been designed for each event-triggered subsystem based on the prediction difference. The dual-model model predictive scheme is employed and a mixed triggered DMPC algorithm is then proposed. Furthermore, the recursive feasibility and closed-loop stability of the proposed algorithm have been proven. By comparing the performance with the single time-triggered or event-triggered scheme, we illustrate that the trade-off between the overall performance and resource usage is realized with the developed mixed triggered DMPC strategy.

References

1. Pop, T., Eles, P., & Peng, Z. (2002). Holistic scheduling and analysis of mixed time/eventtriggered distributed embedded systems. In *Proceedings of the Tenth International Symposium on Hardware/Software Codesign* (pp. 187-192).
2. Yao, J., Xu, X., & Liu, X. (2016). MixCPS: Mixed time/event-triggered architecture of cyber-physical systems. *Proceedings of the IEEE*, *104*(5), 923-937.
3. Xie, Y., Yan, H., & Pang, Z. (2016). Mixed time-triggered and event-triggered controller for industrial IoT applications. In *2016 IEEE International Conference on Industrial Technology (ICIT)* (pp. 2064-2067). IEEE.
4. Willig, A., Matheus, K., & Wolisz, A. (2005). Wireless technology in industrial networks. *Proceedings of the IEEE*, *93*(6), 1130-1151.
5. Hashimoto, K., Adachi, S., & Dimarogonas, D. V. (2015). Distributed aperiodic model predictive control for multi-agent systems. *IET Control Theory and Applications*, *9*(1), 10-20.
6. Li, H., Yan, W., Shi, Y., & Wang, Y. (2015). Periodic event-triggering in distributed receding horizon control of nonlinear systems. *Systems and Control Letters*, *86*, 16-23.
7. Li, L., Shi, P., Agarwal, R. K., Ahn, C. K., & Xing, W. (2019). Event-triggered model predictive control for multiagent systems with communication constraints. *IEEE Transactions on Systems, Man, and Cybernetics: Systems*, *51*(5), 3304-3316.

8. Wei, Y., Li, S., & Wu, J. (2020). Event-triggered distributed model predictive control with optimal network topology. *International Journal of Robust and Nonlinear Control*, *30* (6), 2186-2203.

9. Zou, Y., Su, X., & Niu, Y. (2017). Mixed time/event-triggered distributed predictive control over wired-wireless networks. *Journal of the Franklin Institute*, *354* (9), 3724-3743.

10. Wang, C., & Ong, C. J. (2010). Distributed model predictive control of dynamically decoupled systems with coupled cost. *Automatica*, *46* (12), 2053-2058.

Chapter 6
Self-Triggered DMPC of Networked Systems

6.1 Introduction

In Chapts. 2,3,4,and 5,DMPC algorithms based on event-triggered computation and communication strategies are introduced for energy-saving [1-5]. However, in general, the controlled NCSs are continuously monitored and the preset triggering conditions are continuously checked. This is undesirable for the controlled systems with high sampling costs, such as concentration measurements in chemical processes [6]. Hence, this chapter considers DMPC with self-triggered control [7-10]. Self-triggered control is proactive, that is, the triggering instants are determined in advance based on the past information. And then sensor nodes are activated only at triggering instants and are in sleep during two successive triggering instants, which is different from the event-triggered control, thereby improving the effectiveness of monitoring and checking.

Self-triggered condition is originally derived from stability analysis of the whole system and is only related to the past information. More recently, a term including triggering instants is incorporated to the cost function, thus control inputs and the triggering instant are simultaneously determined by solving an optimal control problem, which can provide a priori performance. In [11-14], self-triggered MPC algorithms for a single subsystem were proposed to relieve communication cost, but the cost function was designed without explicitly considering it. And then [15] and [16] quantified communication cost as a func-

Y. Zou and S. Li, *Distributed Cooperative Model Predictive Control of Networked Systems*,
https://doi.org/10.1007/978-981-19-6084-0_6

tion of triggering interval and added it to the cost function with control performance. In this case, the tuning effect between communication cost and control performance is not obvious when the controlled system closes to its steady state. Thus in order to let it work for almost the whole control process, [17, 18] used an exponential term to describe communication cost and involved it to the cost function as an attenuation factor. For NCSs composed of several subsystems, it becomes more difficult to design a control strategy with a self-triggered mechanism due to the interaction among subsystems. They were usually designed separately in literature [19-21]. Specifically, an MPC controller for each subsystem was designed without considering the self-triggered communication manner, and then a self-triggered condition was obtained from the analysis of system stability. It is conservative of this separate design method and cannot provide a guaranteed control performance. Therefore, it makes sense to develop a united design method of self-triggered scheme and DMPC control algorithm.

This chapter considers the cooperation control problem of linear NCSs with dynamically decoupled subsystems and presents a co-design scheme of self-triggered methodology and DMPC algorithm. The following are its three characteristics. Firstly, not only control performance but also communication cost are explicitly quantified in the cost function, which contributes to achieving a better trade-off between control performance and communication cost. Secondly, in order to reduce communication load, only the first element of the solved control input sequence is applied to the subsystem and sent along with the current state not the whole predictive trajectory to its neighbors for cooperation. Furthermore, stability of the whole system is analyzed and sufficient conditions on design parameters are obtained. The main results of this chapter have been published in [22].

The remainder of this chapter is organized as follows. Section 6. 2 describes the control problem. Section 6. 3 presents the formulation of the self-triggered DMPC problem and expresses its minimization. In Section 6. 4, stability analysis of the whole system is carried out. Section 6. 5 provides the position control of a multi-agent system to verify the effectiveness of the proposed self-triggered DMPC algorithm. Finally, the conclusions are given in

Section 6.6.

Notations: A set $\{1,\ldots,M\}\subset\mathbb{N}$ is denoted as $\mathcal{M}$. The short-hand $\boldsymbol{x}=(x_1,\ldots,x_s)$ denotes a column vector with s components $x_1,\ldots,x_s$. The block-diagonal matrix $\boldsymbol{X}$ with blocks $\boldsymbol{X}_i, i\in\mathcal{M}$, is denoted by $\boldsymbol{X}=\operatorname{diag}(X_1,\ldots,X_M)$. A positive definite matrix $\boldsymbol{P}$ is denoted by $\boldsymbol{P}\succ 0$, its maximum and minimum eigenvalues are denoted by $\overline{\lambda}(\boldsymbol{P})$ and $\underline{\lambda}(\boldsymbol{P})$ respectively. The superscript "T" represents the transposition of vectors or matrices.

6.2 Problem Formulation

Consider a linear NCSs composed of M discrete-time subsystems, each of which is described as

$$\boldsymbol{x}_i(k+1)=\boldsymbol{A}_i\boldsymbol{x}_i(k)+\boldsymbol{B}_i\boldsymbol{u}_i(k),\ \forall i\in\mathcal{M} \tag{6.1}$$

where $\boldsymbol{x}_i(k)\in\mathbb{R}^{m_i}$ is the state of subsystem i at time step k, and $\boldsymbol{u}_i(k)\in\mathbb{R}^{n_i}$ is its control input. $\boldsymbol{A}_i\in\mathbb{R}^{m_i\times m_i}$ and $\boldsymbol{B}_i\in\mathbb{R}^{m_i\times n_i}$ are the system matrix and the input matrix, respectively. Assume that the pair $(\boldsymbol{A}_i,\boldsymbol{B}_i)$, $\forall i\in\mathcal{M}$ is controllable.

Based on (6.1), the whole system can be expressed as

$$\boldsymbol{x}(k+1)=\boldsymbol{A}\boldsymbol{x}(k)+\boldsymbol{B}\boldsymbol{u}(k) \tag{6.2}$$

where $\boldsymbol{x}=(x_1,\ldots,x_M)$, $\boldsymbol{u}=(u_1,\ldots,u_M)$, $\boldsymbol{A}=\operatorname{diag}(A_1,\ldots,A_M)$ and $\boldsymbol{B}=\operatorname{diag}(B_1,\ldots,B_M)$.

For the system in (6.2), we aim at designing a self-triggered DMPC algorithm to decide when and how its control input is updated, thereby achieving the desired cooperation task while using the computation and communication resources in an effective way. Here, we will jointly design the self-triggered mechanism and DMPC controller, and the main idea is: At the r th ($r\in\mathbb{N}$) triggering instant of subsystem $i\in\mathcal{M}$, denoted by t_i^r, its controller solves an optimization control problem over the time horizon $[t_i^r, t_i^r+N_i]$, where $N_i\in\mathbb{N}$ stands for the control horizon of subsystem i, and thus determining its control input sequence $\boldsymbol{U}_i^*=\{u_i^*(t_i^r|t_i^r),\ldots,u_i^*(t_i^r+N_i-1|t_i^r)\}$ and the next triggering instant t_i^{r+1} at the same time. Then during the period of $k\in[t_i^r, t_i^{r+1})$, $\boldsymbol{u}_i^*(t_i^r|t_i^r)$ is applied to the plant and is broadcast to its neighbors with

$\boldsymbol{x}_i(t_i^r)$. The next round of the above computation and communication will be executed at time t_i^{r+1}.

6.3 Co-Design of Self-Triggered Mechanism and DMPC Strategy

6.3.1 *Self-Triggered DMPC Optimization Control Problem*

In self-triggered MPC for a single subsystem, a function which explicitly quantified both control performance and communication cost was constructed in [17, 18]. With the consideration of the interaction between subsystems in (6.2), we will redefine a cost function for it in the following. Specifically, for each subsystem $i \in \mathcal{M}$, its objective function optimized at any triggering instant t_i^r is:

$$J_i[\boldsymbol{x}_i(t_i^r), \boldsymbol{U}_i(t_i^r), T_i] = e^{-\gamma_i T_i} J_i^c[\boldsymbol{x}_i(t_i^r), \boldsymbol{U}_i(t_i^r)] \tag{6.3}$$

where $T_i = t_i^{r+1} - t_i^r$ is defined as the triggering interval and is optimized online together with control input sequence $\boldsymbol{U}_i(t_i^r)$; $e^{-\gamma_i T_i}$ is used to reflect communication cost, and it is defined based on the fact that, as the communication interval T_i decreases, control performance generally becomes better at the expense of communication cost; $\gamma_i \in [0, 1]$ plays a tuning role between communication cost and control performance cost $J_i^c[\boldsymbol{x}_i(t_i^r), \boldsymbol{U}_i(t_i^r)]$. For the system in (6.2), since the subsystems are dynamically decoupled from each other, $J_i^c[\boldsymbol{x}_i(t_i^r), \boldsymbol{U}_i(t_i^r)]$ is usually defined by a function with state coupling in order to reflect the cooperation objectives. Hence, $J_i^c[\boldsymbol{x}_i(t_i^r), \boldsymbol{U}_i(t_i^r)]$ here is written as:

$$\begin{aligned} & J_i^c[\boldsymbol{x}_i(t_i^r), \boldsymbol{U}_i(t_i^r)] \\ = & \sum_{l=0}^{N_i-1} \left(\|\boldsymbol{x}_i(t_i^r + l \mid t_i^r)\|_{\boldsymbol{Q}_i}^2 + \|\boldsymbol{u}_i(t_i^r + l \mid t_i^r)\|_{\boldsymbol{R}_i}^2 \right) \\ & + \sum_{j \in \mathcal{N}_i} \sum_{l=0}^{N_i-1} \|\boldsymbol{x}_i(t_i^r + l \mid t_i^r) - \boldsymbol{x}_j(t_i^r + l \mid t_i^r)\|_{\boldsymbol{Q}_{ij}}^2 \\ & + \|\boldsymbol{x}_i(t_i^r + N_i \mid t_i^r)\|_{\boldsymbol{P}_i}^2 \end{aligned} \tag{6.4}$$

where $(t_i^r+l|t_i^r)$ stands for the prediction of future instant t_i^r+l at t_i^r; $\boldsymbol{Q}_i$, $\boldsymbol{R}_i$, $\boldsymbol{Q}_{ij}$ and $\boldsymbol{P}_i$ are positive definite matrices; $\mathcal{N}_i$ is the set of all subscripts of its neighbors of subsystem i, who can send it information at their triggering instants.

In (6.4), the cooperative objective, expressed by $\sum_{j\in\mathcal{N}_i}\sum_{l=0}^{N_i-1}\|\boldsymbol{x}_i(k+l|k)-\boldsymbol{x}_j(k+l|k)\|_{Q_{ij}}^2$, is added to the local stage cost, such that the local state could be driven to make it approach to its neighbor', thus achieving a designed consensus task. Note that (6.4) can be also adopted for formation control of NCSs in (6.2) through certain state transformation. More precisely, let $\tilde{\boldsymbol{x}}_i=\boldsymbol{x}_i-\boldsymbol{x}_{d_i}$, $\tilde{\boldsymbol{x}}_j=\boldsymbol{x}_j-\boldsymbol{x}_{d_j}$, where $\boldsymbol{x}_{d_i}$ and $\boldsymbol{x}_{d_j}$ represent the desired trajectories of subsystem i and subsystem j respectively, then substitute $\tilde{\boldsymbol{x}}_i$ for $\boldsymbol{x}_i$, and $\boldsymbol{x}_j$ for $\boldsymbol{x}_j$ in (6.4).

Assumption 6.1 $\mathcal{N}_i$ is not empty for any subsystem $i\in\mathcal{M}$. That is, each subsystem can receive information from at least one other subsystem.

In the implementation, the control input $\boldsymbol{u}_i^*(t_i^r|t_i^r)$ acts on the system during the period between two successive triggering instants t_i^r and t_i^{r+1}, i.e. $\boldsymbol{u}_i(k)=\boldsymbol{u}_i^*(t_i^r|t_i^r)$, $k\in[t_i^r,t_i^{r+1})$, and is updated at t_i^{r+1}, hence certain constraints on control are imposed as follows,

$$\boldsymbol{u}_i(t_i^r+l|t_i^r)=\begin{cases}\boldsymbol{u}_i(t_i^r|t_i^r), 0\leqslant l<T_i,\\ \boldsymbol{u}_i(t_i^r+T_i|t_i^r), T_i\leqslant l<2T_i,\\ \vdots\\ \boldsymbol{u}_i(t_i^r+(\eta-1)T_i|t_i^r), (\eta-1)T_i\leqslant l<\eta T_i\end{cases} \tag{6.5}$$

where $\eta\in\mathbb{N}$ and the prediction horizon $N_i=\eta T_i$ varies with T_i. The blocking control constraint in (6.5) is favorable for decreasing computation load by reducing the number of decision variables in the optimized control problem, and also for achieving a better control performance by tuning η.

Remark 6.1 In (6.3), not only control performance but also communication cost are explicitly reflected, and different trade-offs between them can be obtained by tuning γ_i. For $\gamma_i=0$, the exponential term $e^{-\gamma_i T_i}$ becomes a constant 1, and only control performance is considered in (6.3) which is the same as the traditional cost function. For $\gamma_i\in(0,1]$, the positive term $e^{-\gamma_i T_i}$ is less than 1. In this case, both communication cost and control performance are considered in the cost function. Further analysis found that as T_i increases, $e^{-\gamma_i T_i}$ decreases dramatically, which indicates that communication cost will be dominated in optimization and a longer triggering interval will be obtained. This point will be illustratively displayed in Section 6.5. As for the construction of (6.3), firstly choose matrices $\boldsymbol{Q}_i$, $\boldsymbol{Q}_{ij}$, $\boldsymbol{R}_i$ and $\boldsymbol{P}_i$ to obtain the optimal level of control performance, and then regulate γ_i to guarantee a certain desired communication behavior and control performance level.

For NCSs in (6.2), the dynamically decoupled subsystems are required to exchange information with their neighbors to achieve the assigned cooperation tasks. But for each subsystem i, the future state trajectories of its neighbors, i.e. $\boldsymbol{x}_j(t_i^r+l\,|t_i^r)$, $j\in\mathcal{N}_i$ in (6.3), are unknown because only $\boldsymbol{u}_j^*(t_j^{c_i}\,|t_j^{c_i})$ and $\boldsymbol{x}_j(t_j^{c_i})$ are available at t_i^r, where $t_j^{c_i}$, $c_i\in\mathbb{N}$ is the most recent triggering instant of subsystem j from t_i^r. Hence, some assumed trajectories have to be made for all neighbors prior to each update. In this chapter, the assumed state trajectory of each neighbor subsystem $j\in\mathcal{N}_i$ is generated by:

$$\hat{\boldsymbol{x}}_j(t_i^r+sT_i+l\,|t_i^r)=\mu_j\hat{\boldsymbol{x}}_j(t_i^r+sT_i\,|t_i^r) \tag{6.6}$$

where $s=0,\ldots,\eta-1$, $l=1,\ldots,T_i$, and μ_j is a design parameter and decides which kind of information of its neighbor j is used by subsystem i for cooperation, which directly influences system stability and convergence rate. Sufficient conditions about μ_j will be further constructed to ensure stability in Section 6.4. For $s=0$, the initial assumed state $\hat{\boldsymbol{x}}_j(t_i^r\,|t_i^r)$ is defined by:

$$\hat{\boldsymbol{x}}_j(t_i^r\,|t_i^r)=\boldsymbol{A}_j^{t_i^r-t_j^{c_i}}\boldsymbol{x}_j(t_j^{c_i})+\boldsymbol{B}_j^{(t_i^r-t_j^{c_i}-1)}\boldsymbol{u}_j(t_j^{c_i}) \tag{6.7}$$

where $\boldsymbol{B}_j^{(l)}=\sum_{h=0}^{l}\boldsymbol{A}_j^h\boldsymbol{B}_j$.

Remark 6.2 For each subsystem $i\in\mathcal{M}$, the actual state of its neighbor $j\in\mathcal{N}_i$ at t_i^r is:

$$\boldsymbol{x}_j(t_i^r)=\boldsymbol{A}_j^{t_i^r-t_j^{c_i}}\boldsymbol{x}_j(t_j^{c_i})+\boldsymbol{B}_j^{(t_i^r-t_j^{c_i}-1)}\boldsymbol{u}_j(t_j^{c_i}) \tag{6.8}$$

It is obtained from (6.7) and (6.8) that the assumed state of neighbor j at t_i^r

is equal to its actual state, i. e. $\hat{\boldsymbol{x}}_j(t_i^r|t_i^r)=\boldsymbol{x}_j(t_i^r)$. However, there always exists deviation between $\hat{\boldsymbol{x}}_j(t_i^r+sT_i+l)$ and $\boldsymbol{x}_j(t_i^r+sT_i+l)$ for all $s=0,\ldots,\eta-1$ and $l=1,\ldots,T_i$, since the actual state of neighbor j at future time $\boldsymbol{x}_j(t_i^r+sT_i+l)=\boldsymbol{A}_j\boldsymbol{x}_j(t_i^r+sT_i+l-1)+\boldsymbol{B}_j\boldsymbol{u}_j(t_i^r+sT_i+l-1)$, where the actual control input $\boldsymbol{u}_j(t_i^r+sT_i+l-1)$ is yet not available for subsystem i at t_i^r, and a corresponding assumed state is generated according to (6.6), i. e. $\hat{\boldsymbol{x}}_j(t_i^r+sT_i+l)$.

In summarize, the optimization control problem of the proposed DMPC based on self-triggered mechanism can be formulated in the following. For each subsystem i, given its local state measurement $\boldsymbol{x}_i(t_i^r)$, updating the latest received information $\boldsymbol{x}_j(t_j^{c_i})$ and $\boldsymbol{u}_j^*(t_j^{c_i}|t_j^{c_i})$, subsystem i solves *Problem* P_i online at t_i^r to determine t_i^{r+1} and $\boldsymbol{u}_i^*(t_i^r|t_i^r)$.

Problem $\boldsymbol{P}_i$:

$$\min_{T_i,\boldsymbol{U}_i(t_i^r)} J_i[\boldsymbol{x}_i(t_i^r),\boldsymbol{U}_i(t_i^r),T_i] \tag{6.9}$$

subject to:

$$\boldsymbol{x}_i(t_i^r+l+1|t_i^r)=\boldsymbol{A}_i\boldsymbol{x}_i(t_i^r+l|t_i^r)+\boldsymbol{B}_i\boldsymbol{u}_i(t_i^r+l|t_i^r) \tag{6.10}$$

$$\hat{\boldsymbol{x}}_j(t_i^r+sT_i+l+1|t_i^r)=\mu_j\hat{\boldsymbol{x}}_j(t_i^r+sT_i|t_i^r) \tag{6.11}$$

control constraints (6.5),

$$T_i\in\{1,\ldots,\overline{T}_i\} \tag{6.12}$$

where $s=0,\ldots,\eta-1$; $l=0,\ldots,\eta T_i-1$ and $\overline{T}_i\in\mathbb{N}$ is an upper bound of T_i. Due to the fact that local subsystems may deviate from their cooperation objective if there is no communication among them for a long time, constraint (6.12) is imposed to ensure control performance, and certain conditions on the choice of $\overline{T}_i\in\mathbb{N}$ will be given later for stability guarantees.

6.3.2 Explicit Solution of the Self-Triggered MPC

In this subsection, the formulated *Problem* $\boldsymbol{P}_i$ will be solved and an explicit solution will be presented.

Firstly, the predicted state of subsystem i at future time $t_i^r+sT_i+l|t_i^r$

for any $s=0,\ldots,\eta-1$ and $l=1,\ldots,T_i$ is rewritten as follows based on the prediction model (6.10) and the imposed control constraints (6.5),

$$x_i(t_i^r+sT_i+l\,|t_i^r)=\mathbf{A}_i^l x_i(t_i^r+sT_i\,|t_i^r)+B_i^{(l-1)}u_i(t_i^r+sT_i\,|t_i^r) \quad (6.13)$$

The substitution of Eqs. (6.11) and (6.13) into *Problem* $\mathbf{P}_i$ yields that

$$\begin{aligned}
&J_i[\boldsymbol{x}_i(t_i^r),\boldsymbol{U}_i(t_i^r),T_i]\\
=&e^{-\gamma_i T_i}\times\Bigg\{\sum_{s=0}^{\eta-1}\Big[\|\boldsymbol{x}_i(t_i^r+sT_i\,|t_i^r)\|^2_{\widetilde{\boldsymbol{Q}}_{ij}}+\|\boldsymbol{u}_i(t_i^r+sT_i\,|t_i^r)\|^2_{\widetilde{\boldsymbol{R}}_{ij}}\\
&+2\boldsymbol{x}_i^{\mathrm{T}}(t_i^r+sT_i\,|t_i^r)\widetilde{\mathrm{N}}_{ij}\boldsymbol{u}_i(t_i^r+sT_i\,|t_i^r)-2\sum_{j\in\mathcal{N}_i}\mu_j^s\boldsymbol{x}_i^{\mathrm{T}}(t_i^r+sT_i\,|t_i^r)\hat{\boldsymbol{Q}}_{ij}\boldsymbol{x}_j(t_i^r)\\
&-2\sum_{j\in\mathcal{N}_i}\mu_j^s\boldsymbol{u}_i^{\mathrm{T}}(t_i^r+sT_i\,|t_i^r)\overline{\boldsymbol{Q}}_{ij}\boldsymbol{x}_j(t_i^r)+\sum_{j\in\mathcal{N}_i}\mu_j^{2s}T_i\|\boldsymbol{x}_j(t_i^r)\|^2_{\boldsymbol{Q}_{ij}}\Big]\\
&+\|\boldsymbol{x}_i(t_i^r+\eta T_i\,|t_i^r)\|^2_{\mathbf{P}_i}\Bigg\}
\end{aligned} \quad (6.14)$$

where

$$\begin{aligned}
\widetilde{\boldsymbol{Q}}_{ij}&=\sum_{l=0}^{T_i-1}(\boldsymbol{A}_i^l)^{\mathrm{T}}(\boldsymbol{Q}_i+\sum_{j\in\mathcal{N}_i}\boldsymbol{Q}_{ij})\boldsymbol{A}_i^l,\\
\widetilde{\boldsymbol{R}}_{ij}&=T_i\boldsymbol{R}_i+\sum_{l=1}^{T_i-1}[\boldsymbol{B}_i^{(l-1)}]^{\mathrm{T}}(\boldsymbol{Q}_i+\sum_{j\in\mathcal{N}_i}\boldsymbol{Q}_{ij})\boldsymbol{B}_i^{(l-1)},\\
\widetilde{\mathbf{N}}_{ij}&=\sum_{l=1}^{T_i-1}(\boldsymbol{A}_i^l)^{\mathrm{T}}(\boldsymbol{Q}_i+\sum_{j\in\mathcal{N}_i}\boldsymbol{Q}_{ij})\boldsymbol{B}_i^{(l-1)},\\
\hat{\boldsymbol{Q}}_{ij}&=\sum_{l=0}^{T_i-1}\mu_j(\boldsymbol{A}_i^l)^{\mathrm{T}}\boldsymbol{Q}_{ij},\overline{\boldsymbol{Q}}_{ij}=\sum_{l=1}^{T_i-1}\mu_j[\boldsymbol{B}_i^{(l-1)}]^{\mathrm{T}}\boldsymbol{Q}_{ij}.
\end{aligned}$$

Furthermore, the Eq. (6.14) can be converted to a function only related to the triggering interval T_i by using the dynamic programming. More precisely, when $s=\eta-1$, let

$$\frac{\partial J_i[\boldsymbol{x}_i(t_i^r),\boldsymbol{U}_i(t_i^r),T_i]}{\partial\boldsymbol{u}_i[t_i^r+(\eta-1)T_i\,|t_i^r]}=0 \quad (6.15)$$

we get

$$\boldsymbol{u}_i[t_i^r+(\eta-1)T_i\,|t_i^r]=-F_{ii}^{(\eta-1)}\boldsymbol{x}_i[t_i^r+(\eta-1)T_i\,|t_i^r]+\sum_{j\in\mathcal{N}_i}\mu_j^{\eta-1}F_{ij}^{(\eta-1)}\boldsymbol{x}_j(t_i^r) \tag{6.16}$$

where

$$\Phi_{ii}^{(\eta-1)}=\widetilde{\boldsymbol{R}}_{ij}+[\boldsymbol{B}_i^{(T_i-1)}]^{\mathrm{T}}\boldsymbol{P}_i\boldsymbol{B}_i^{(T_i-1)},$$

$$F_{ij}^{(\eta-1)}=[\Phi_{ii}^{(\eta-1)}]^{-1}\overline{\boldsymbol{Q}}_{ij},$$

$$F_{ii}^{(\eta-1)}=[\Phi_{ii}^{(\eta-1)}]^{-1}[\widetilde{N}_{ij}+(\boldsymbol{A}_i^{T_i})^{\mathrm{T}}\boldsymbol{P}_i\boldsymbol{B}_i^{(T_i-1)}]^{\mathrm{T}}.$$

When $s=\eta-2$, substituting (6.16) and (6.13) into (6.14), and then let

$$\frac{\partial J_i[\boldsymbol{x}_i(t_i^r),\boldsymbol{U}_i(t_i^r),T_i]}{\partial\boldsymbol{u}_i[t_i^r+(\eta-2)T_i\,|t_i^r]}=0 \tag{6.17}$$

we have

$$\boldsymbol{u}_i[t_i^r+(\eta-2)T_i\,|t_i^r]=-F_{ii}^{(\eta-2)}\boldsymbol{x}_i[k+(\eta-2)T_i\,|t_i^r]+\sum_{j\in\mathcal{N}_i}\mu_j^{\eta-2}F_{ij}^{(\eta-2)}\boldsymbol{x}_j(t_i^r) \tag{6.18}$$

where

$$\boldsymbol{P}_i^{(\eta-1)}=\widetilde{\boldsymbol{Q}}_{ij}+(\boldsymbol{A}_i^{T_i})^{\mathrm{T}}\boldsymbol{P}_i\boldsymbol{A}_i^{T_i}-[\widetilde{\boldsymbol{N}}_{ij}+(\boldsymbol{A}_i^{T_i})^{\mathrm{T}}\boldsymbol{P}_i\boldsymbol{B}_i^{(T_i-1)}]F_{ii}^{(\eta)},$$

$$\Lambda_{ij}^{(\eta-1)}=[F_{ii}^{(\eta-1)})^{\mathrm{T}}\overline{\boldsymbol{Q}}_{ij}-\hat{\boldsymbol{Q}}_{ij},$$

$$\Phi_{ii}^{(\eta-2)}=\widetilde{\boldsymbol{R}}_{ij}+[\boldsymbol{B}_i^{(T_i-1)}]^{\mathrm{T}}\boldsymbol{P}_i^{(\eta-1)}\boldsymbol{B}_i^{(T_i-1)},$$

$$F_{ii}^{(\eta-2)}=[\Phi_{ii}^{(\eta-2)}]^{-1}[\widetilde{\boldsymbol{N}}_{ij}+(\boldsymbol{A}_i^{T_i})^{\mathrm{T}}\boldsymbol{P}_i^{(\eta-1)}\boldsymbol{B}_i^{(T_i-1)}]^{\mathrm{T}},$$

$$F_{ij}^{(\eta-2)}=[\Phi_{ii}^{(\eta-2)}]^{-1}\{\overline{\boldsymbol{Q}}_{ij}-\mu_j[\boldsymbol{B}_i^{(T_i-1)}]^{\mathrm{T}}\Lambda_{ij}^{(\eta-1)}\}.$$

When $s=\eta-3,\ldots,0$, with a similar procedure, we get

$$\boldsymbol{u}_i(t_i^r+sT_i\,|t_i^r)=-F_{ii}^{(s)}\boldsymbol{x}_i(t_i^r+sT_i\,|t_i^r)+\sum_{j\in\mathcal{N}_i}\mu_j^sF_{ij}^{(s)}\boldsymbol{x}_j(t_i^r) \tag{6.19}$$

where

$$\boldsymbol{P}_i^{(s+1)}=\widetilde{\boldsymbol{Q}}_{ij}+(\boldsymbol{A}_i^{T_i})^{\mathrm{T}}\boldsymbol{P}_i^{(s+2)}\boldsymbol{A}_i^{T_i}-[\widetilde{N}_{ij}+(\boldsymbol{A}_i^{T_i})^{\mathrm{T}}\boldsymbol{P}_i^{(s+2)}\boldsymbol{B}_i^{(T_i-1)}]F_{ii}^{(s+1)},$$

$$\Lambda_{ij}^{(s+1)}=[F_{ii}^{(s+1)}]^{\mathrm{T}}\overline{\boldsymbol{Q}}_{ij}-\hat{\boldsymbol{Q}}_{ij}+\mu_j[\boldsymbol{A}_i^{T_i}-\boldsymbol{B}_i^{(T_i-1)}F_{ii}^{(s+1)}]^{\mathrm{T}}\Lambda_{ij}^{(s+2)},$$

$$\Phi_{ii}^{(s)}=\widetilde{R}_{ij}+[\boldsymbol{B}_i^{(T_i-1)}]^{\mathrm{T}}\boldsymbol{P}_i^{(s+1)}\boldsymbol{B}_i^{(T_i-1)},$$

$$F_{ii}^{(s)}=[\Phi_{ii}^{(s)}]^{-1}[\widetilde{\boldsymbol{N}}_{ij}+(\boldsymbol{A}_i^{T_i})^{\mathrm{T}}\boldsymbol{P}_i^{(s+1)}\boldsymbol{B}_i^{(T_i-1)}]^{\mathrm{T}},$$

$$F_{ij}^{(s)}=[\Phi_{ii}^{(s)}]^{-1}\{\overline{\boldsymbol{Q}}_{ij}-[\boldsymbol{B}_i^{(T_i-1)}]^{\mathrm{T}}\Lambda_{ij}^{(s+1)}\boldsymbol{A}_j^{T_i}\}.$$

Substitute Eqs. (6.13), (6.16), (6.18) and all the equations in (6.19) into (6.14), we then get the following *Problem* $\widetilde{\boldsymbol{P}}_i$, which is equal to *Problem* $\boldsymbol{P}_i$ and is one of the main results of this subsection.

Problem $\widetilde{\boldsymbol{P}}_i$:

$$\min_{T_i} e^{-\gamma_i T_i}\left[C_i^{(0)}+\|\boldsymbol{x}_i(t_i^r)\|_{\boldsymbol{P}_i^{(0)}}^2+2\sum_{j\in\mathcal{N}_i}\boldsymbol{x}_i^{\mathrm{T}}(t_i^r)\Lambda_{ij}^{(0)}\boldsymbol{x}_j(t_i^r)\right] \tag{6.20}$$

with $T_i\in\{1,\ldots,\overline{T}_i\}$, where

$$C_i^{(\eta-1)}=\sum_{j\in\mathcal{N}_i}\mu_j^{2\eta}T_i\|\boldsymbol{x}_j(t_i^r)\|_{\overline{Q}_{ij}}^2-\left\|\sum_{j\in\mathcal{N}_i}\mu_j^{\eta}F_{ij}^{(\eta-1)}\boldsymbol{x}_j(t_i^r)\right\|_{\Phi_{ii}^{(\eta-1)}}^2,$$

$$C_i^{(s)}=C_i^{(s+1)}+\sum_{j\in\mathcal{N}_i}\mu_j^{2(s+1)}T_i\|\boldsymbol{x}_j(t_i^r)\|_{\overline{Q}_{ij}}^2-\left\|\sum_{j\in\mathcal{N}_i}\mu_j^{s+1}F_{ij}^{(s)}\boldsymbol{x}_j(t_i^r)\right\|_{\boldsymbol{\Phi}_{ii}^{(s)}}^2,$$

$\forall s=0,\ldots,\eta-2.$

Remark 6.3 The control input for any time instant during two successive triggering instants satisfies

$$\boldsymbol{u}_i(k)=\boldsymbol{u}_i^*(t_i^r|t_i^r)=-F_{ii}^{(0)}\boldsymbol{x}_i(t_i^r)+\sum_{j\in\mathcal{N}_i}F_{ij}^{(0)}\boldsymbol{x}_j(t_i^r) \tag{6.21}$$

where $k\in[t_i^r,t_i^{r+1})$, feedback gains $F_{ii}^{(0)}$ and $F_{ij}^{(0)}$ are the ones corresponding to the optimal triggering interval T_i^*. In the implementation, $\boldsymbol{P}_i^{(0)}$, $\Lambda_{ij}^{(0)}$, $F_{ii}^{(0)}$, $F_{ij}^{(s)}$ and $\Phi_{ii}^{(s)}$ for all $T_i\in\{1,\ldots,\overline{T}_i\}$, $s=0,\ldots,\eta-1$ are calculated and stored offline so as to relieve the online computation burden. Then during the online runtime, each controller $i\in\mathcal{M}$ receives state measurement $\boldsymbol{x}_i(t_i^r)$ at its own triggering instant t_i^r, and computes (6.19) in *Problem* $\widetilde{\boldsymbol{P}}_i$. Subsequently, T_i^* is determined by minimizing the cost function, then the next triggering instant is obtained, i.e., $t_i^{r+1}=t_i^r+T_i^*$, and the control input $u_i(k)$, $\forall k\in[t_i^r,t_i^{r+1})$ is determined according to (6.21).

6.3.3 Self-Triggered DMPC Algorithm

In this subsection, the self-triggered DMPC algorithm is presented and its the main steps are summarized in Algorithm 1.

Algorithm 1 Self-triggered DMPC

Off-line: For all subsystem $i \in \mathcal{M}$, choose η, $\overline{T}_i$, $\boldsymbol{Q}_i$, $\boldsymbol{R}_i$, $\boldsymbol{P}_i$, $\boldsymbol{Q}_{ij}$, μ_j, $\forall j \in \mathcal{N}_i$; For any $T_i \in \{1, \ldots, \overline{T}_i\}$, compute and store $\boldsymbol{P}_i^{(0)}$, $\Lambda_{ij}^{(0)}$, $F_{ii}^{(0)}$, $F_{ij}^{(s)}$ and $\Phi_{ii}^{(s)}$, $\forall s = 0, \ldots, \eta - 1$.

On-line:

Initialization. For any subsystem $i \in \mathcal{M}$, set $t_i^0 = 0$ and running step t_{run}.
for $k = 0, 1, \ldots, t_{run}$ **do**
if $k = t_i^r$ **then**
    Sample and send $\boldsymbol{x}_i(t_i^r)$ to controller i;
    Update $\hat{\boldsymbol{x}}_j(t_i^r + l \mid t_i^r)$, $\forall j \in \mathcal{N}_i$ based on (6.6);
    Solve (6.20) to obtain T_i^*;
    Substitute the corresponding $F_{ii}^{(0)}$ and $F_{ij}^{(0)}$ to (6.21) and calculate $\boldsymbol{u}_i^*(t_i^r \mid t_i^r)$;
    Send $\boldsymbol{u}_i^*(t_i^r \mid t_i^r)$ to the actuator and apply it;
    Send $\boldsymbol{u}_i^*(t_i^r \mid t_i^r)$ and $\boldsymbol{x}_i(t_i^r)$ to all subsystem $j \in \mathcal{N}_i$;
    Let $t_i^{r+1} = t_i^r + T_i^*$, $r \leftarrow r+1$.
else
    $\boldsymbol{u}_i(k) = \boldsymbol{u}_i^*(t_i^r \mid t_i^r)$, $k \in [t_i^r, t_i^{r+1})$ is applied.
end if
end for

Remark 6.4 It is found that each controller i in Algorithm 1 requires to update its neighbors' assumed state trajectories at its triggering instant t_i^r, thus a buffer is needed to record the information received from its neighbors, note that only the latest received information from the same neighbor is kept according to (6.6) and (6.7). Besides, controller i won't be activated to update the control law immediately even if subsystem i gets new data from its neighbors at $k \in (t_i^r, t_i^{r+1})$. Instead, it reads the information from the buffer until t_i^{r+1} to optimize control input and triggering instant.

6.4 Stability Analysis

In this section, sufficient conditions on the design parameters will be constructed by stability analysis of the overall closed-loop system.

Theorem 6.1 *Suppose that* $(\boldsymbol{A}_i^{T_i}, \boldsymbol{B}_i^{T_i})$, $\forall T_i \in \{1, \ldots, \overline{T}_i\}$ *is controllable and Assumption 6.1 holds. The controlled system in* (6.2) *is stabilizing if* μ_j *and the terminal matrix* P_i *are selected such that* $\mathbf{A}_{\tilde{\mathcal{N}}_i}(T_i)$ *satisfies the condition* (6.22).

$$\boldsymbol{A}_{\widetilde{\mathcal{N}}_i}^{\mathrm{T}}(T_i)\boldsymbol{P}_i\boldsymbol{A}_{\widetilde{\mathcal{N}}_i}(T_i)-\boldsymbol{W}_i^{\mathrm{T}}\boldsymbol{P}_i\boldsymbol{W}_i<0,\ \forall\, T_i\in\{1,\dots,\overline{T}_i\} \tag{6.22}$$

where $\boldsymbol{A}_{\widetilde{\mathcal{N}}_i}(T_i)$ *is an expanded matrix including elements* $\boldsymbol{A}_i^{T_i}-\boldsymbol{B}_i^{(T_i-1)}F_{ii}^{(0)}$ *and* $\boldsymbol{B}_i^{(T_i-1)}F_{ij}^{(0)}$, $\forall j\in\mathcal{N}_i$. *More clearly, we number the subsystems in* $\mathcal{N}_i$ *as* $j_1,\dots,j_w$ *and* $j_1<\dots<i<\dots<j_w$. *Then* $\boldsymbol{A}_{\widetilde{\mathcal{N}}_i}(T_i)$ *is defined by* $\boldsymbol{A}_{\widetilde{\mathcal{N}}_i}(T_i)=[\boldsymbol{B}_i^{(T_i-1)}F_{ij_1}^{(0)},\dots,\boldsymbol{A}_i^{T_i}-\boldsymbol{B}_i^{(T_i-1)}F_{ii}^{(0)},\dots,\boldsymbol{B}_i^{(T_i-1)}F_{ij_w}^{(0)}]$. *Besides,* $\boldsymbol{W}_i\in\{0,1\}^{m_i\times m_{\widetilde{\mathcal{N}}_i}}$ *is a transformation matrix that makes* $\boldsymbol{x}_i=\boldsymbol{W}_i\boldsymbol{x}_{\widetilde{\mathcal{N}}_i}$ *hold, where* $\boldsymbol{x}_{\widetilde{\mathcal{N}}_i}=(x_{ij_1},\dots,x_i,\dots,x_{ij_w})$, $m_{\widetilde{\mathcal{N}}_i}=m_i+\sum_{h=1}^{w}m_{jh}$

Proof Through Algorithm 1, the dynamically decoupled subsystems in (6.2) asynchronously and aperiodically update control law and broadcast system information. For each subsystem $i\in\mathcal{M}$, denote its own self-triggered time sequence as $\{t_i^r, r\in\mathbb{N}\}$, and generally $t_i^r\neq t_j^r$, $\forall\, i,j\in\mathcal{M}$ and $i\neq j$. Prior to analyzing stability of the whole system, we first combine and rank all the triggering instants of M subsystems in ascending order, then denote it as $\{\tau_t, t\in\mathbb{N}\}$. Furthermore a Lyapunov candidate function of the closed-loop system is defined as $\boldsymbol{V}[x(\tau_t)]=\sum_{i\in\mathcal{M}}\boldsymbol{V}_i[\boldsymbol{x}_i(\tau_t)]$, where $\boldsymbol{V}_i[x_i(\tau_t)]=\boldsymbol{x}_i^{\mathrm{T}}(\tau_t)\boldsymbol{P}_i\boldsymbol{x}_i(\tau_t)$. In the following, we will prove that (i) $\boldsymbol{V}[\boldsymbol{x}(\tau_t)]<\overline{\boldsymbol{V}}[\boldsymbol{x}(\tau_t)]$, where $\overline{\boldsymbol{V}}[\boldsymbol{x}(\tau_t)]=\sum_{i\in\mathcal{M}}\boldsymbol{V}_i[\boldsymbol{x}_i(t_i^{[\tau_t]})]>0$, $t_i^{[\tau_t]}=\arg\min_{t_i^r<\tau_t}\tau_t-t_i^r$; and (ii) $\lim_{\tau_t\to\infty}\boldsymbol{V}[\boldsymbol{x}(\tau_t)]=0$.

(i) Based on (6.13) and (6.21), the state of subsystem $i\in\mathcal{M}$ at any time instant τ_t, $t\in\mathbb{N}$ is obtained by:

$$\begin{aligned}
&\boldsymbol{x}_i(\tau_t)\\
&=\boldsymbol{A}_i^{\Delta_t}\boldsymbol{x}_i(t_i^{[\tau_t]})+\boldsymbol{B}_i^{(\Delta_t-1)}\times[-F_{ii}^{(0)}\boldsymbol{x}_i(t_i^{[\tau_t]})+\sum_{j\in\mathcal{N}_i}F_{ij}^{(0)}\boldsymbol{x}_j(t_i^{[\tau_t]})]\\
&=[\boldsymbol{A}_i^{\Delta_t}-\boldsymbol{B}_i^{(\Delta_t-1)}F_{ii}^{(0)}]\boldsymbol{x}_i(t_i^{[\tau_t]})+\sum_{j\in\mathcal{N}_i}\boldsymbol{B}_i^{(\Delta_t-1)}F_{ij}^{(0)}\boldsymbol{x}_j(t_i^{[\tau_t]})\\
&=\boldsymbol{A}_{\widetilde{\mathcal{N}}_i}(\Delta_t)\boldsymbol{x}_{\widetilde{\mathcal{N}}_i}(t_i^{[\tau_t]})
\end{aligned} \tag{6.23}$$

where $\Delta_t=\tau_t-t_i^{[\tau_t]}\in\{1,\dots,\overline{T}_i\}$. Then the difference between $\boldsymbol{V}_i[\boldsymbol{x}_i(\tau_t)]$ and $\boldsymbol{V}_i[\boldsymbol{x}_i(t_i^{[\tau_t]})]$ is obtained through (6.23) and state transformation $\boldsymbol{x}_i=\boldsymbol{W}_i\boldsymbol{x}_{\widetilde{\mathcal{N}}_i}$:

$$\begin{aligned}&\boldsymbol{V}_i[\boldsymbol{x}_i(\tau_t)]-\boldsymbol{V}_i[\boldsymbol{x}_i(t_i^{[\tau_t]})]\\&=\boldsymbol{x}_{\widetilde{\mathcal{N}}_i}^{\mathrm{T}}(t_i^{[\tau_t]})\boldsymbol{A}_{\widetilde{\mathcal{N}}_i}^{\mathrm{T}}(\Delta_t)\boldsymbol{P}_i\boldsymbol{A}_{\widetilde{\mathcal{N}}_i}(\Delta_t)\boldsymbol{x}_{\widetilde{\mathcal{N}}_i}(t_i^{[\tau_t]})-\boldsymbol{x}_i^{\mathrm{T}}(t_i^{[\tau_t]})\boldsymbol{P}_i\boldsymbol{x}_i(t_i^{[\tau_t]})\\&=\boldsymbol{x}_{\widetilde{\mathcal{N}}_i}^{\mathrm{T}}(t_i^{[\tau_t]})[\boldsymbol{A}_{\widetilde{\mathcal{N}}_i}^{\mathrm{T}}(\Delta_t)\boldsymbol{P}_i\boldsymbol{A}_{\widetilde{\mathcal{N}}_i}(\Delta_t)-\boldsymbol{W}_i^{\mathrm{T}}\boldsymbol{P}_i\boldsymbol{W}_i]\boldsymbol{x}_{\widetilde{\mathcal{N}}_i}(t_i^{[\tau_t]})\end{aligned}\tag{6.24}$$

Due to the satisfaction of condition (6.22), it holds that $\boldsymbol{A}_{\widetilde{\mathcal{N}}_i}^{\mathrm{T}}(\Delta_t)\boldsymbol{P}_i\boldsymbol{A}_{\widetilde{\mathcal{N}}_i}(\Delta_t)-\boldsymbol{W}_i^{\mathrm{T}}\boldsymbol{P}_i\boldsymbol{W}_i<0$, thus

$$\boldsymbol{V}_i[\boldsymbol{x}_i(\tau_t)]-\boldsymbol{V}_i[\boldsymbol{x}_i(t_i^{[\tau_t]})]<0 \tag{6.25}$$

According to (6.25), we get that

$$\boldsymbol{V}[\boldsymbol{x}(\tau_t)]-\overline{\boldsymbol{V}}[\boldsymbol{x}(\tau_t)]=\sum_{i\in\mathcal{M}}\boldsymbol{V}_i[\boldsymbol{x}_i(\tau_t)]-\boldsymbol{V}_i[\boldsymbol{x}_i(t_i^{[\tau_t]})]<0 \tag{6.26}$$

(ii) For each subsystem $i\in\mathcal{M}$, when $t_i^{[\tau_{t+1}]}=t_i^{[\tau_t]}$, it has $\boldsymbol{V}_i[\boldsymbol{x}_i(t_i^{[\tau_{t+1}]})]-\boldsymbol{V}_i[\boldsymbol{x}_i(t_i^{[\tau_t]})]=0$; when $t_i^{[\tau_{t+1}]}\neq t_i^{[\tau_t]}$, that is $t_i^{[\tau_{t+1}]}=\tau_t>t_i^{[\tau_t]}$, it gets from (6.25) that $\boldsymbol{V}_i[x_i(t_i^{[\tau_{t+1}]})]-\boldsymbol{V}_i[x_i(t_i^{[\tau_t]})]=\boldsymbol{V}_i[x_i(\tau_t)]-\boldsymbol{V}_i[x_i(t_i^{[\tau_t]})]<0$. Since there exists at least one subsystem $\zeta\in\mathcal{M}$, which is triggered at time $\tau_t, t\in\mathbb{N}$ and satisfies $\boldsymbol{V}_\zeta[\boldsymbol{x}_\zeta(t_\zeta^{[\tau_{t+1}]})]-\boldsymbol{V}_\zeta[\boldsymbol{x}_\zeta(t_\zeta^{[\tau_t]})]<0$, thus the difference between $\overline{\boldsymbol{V}}[\boldsymbol{x}(\tau_{t+1})]$ and $\overline{\boldsymbol{V}}[\boldsymbol{x}(\tau_t)]$ satisfies

$$\overline{\boldsymbol{V}}[\boldsymbol{x}(\tau_{t+1})]-\overline{\boldsymbol{V}}[\boldsymbol{x}(\tau_t)]=\sum_{i\in\mathcal{M}}\boldsymbol{V}_i[\boldsymbol{x}_i(t_i^{[\tau_{t+1}]})]-\boldsymbol{V}_i[\boldsymbol{x}_i(t_i^{[\tau_t]})]<0 \tag{6.27}$$

Therefore, it holds that $\lim_{\tau_t\to\infty}\overline{V}[\boldsymbol{x}(\tau_t)]=0$. Then following the results in (i) yields

$$0\leqslant\lim_{\tau_t\to\infty}\boldsymbol{V}[\boldsymbol{x}(\tau_t)]\leqslant\lim_{\tau_t\to\infty}\overline{\boldsymbol{V}}[\boldsymbol{x}(\tau_{t+1})]=0 \tag{6.28}$$

From (6.28), it has $\lim_{\tau_t\to\infty}V(x(\tau_t))=0$ and $\lim_{\tau_t\to\infty}x(\tau_t)=0$, which indicates that the controlled system gradually evolves to the origin through Algorithm 1. It completes the proof.

□

6.5 Example

In this section, the angular consensus problem of four single-link robot manipulators as shown in Fig. 6. 1 is considered and simulation on MATLAB through the proposed self-triggered DMPC algorithm is made to illustrate its effectiveness. For each manipulator, the DC motor coupled to the rigid link through a gear train is used to produce the generalized force u_i . By Newton's Second Law, the dynamics model of the robot manipulator i, $i = 1, 2, 3, 4$ is formulated [23]:

$$Z_i \ddot{q}_i + D_i \dot{q}_i + mgLsin(q_i) = \boldsymbol{u}_i \tag{6.29}$$

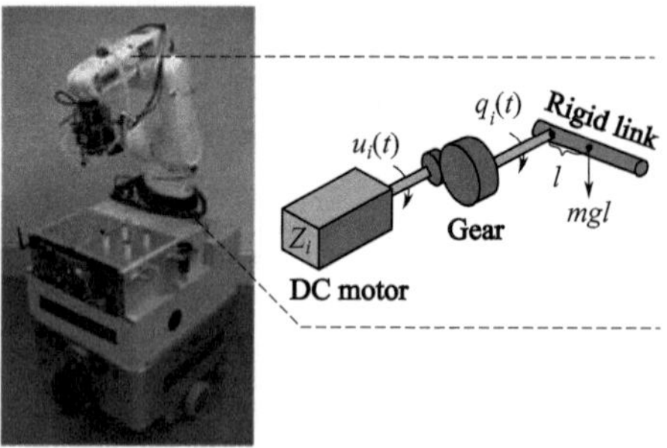

Fig. 6. 1 Single-link robot manipulator i ($i = 1,2,3,4$)

where $\boldsymbol{u}_i$ stands for the control input, q_i is the angle of its rigid link, $\dot{q}_i$ denotes the angular velocity, Z_i is the rotation inertia of the servo motor, D_i is the damping factor, m is the link's total mass, l is denoted as the distance from joint axis to the center of mass, and g is the gravitational acceleration. Assume that the four manipulators have the same physical parameters: $Z_i = 1$, $D_i = 0.002$, and $mgl = 5$.

Let $\boldsymbol{x}_i = (q_i, \dot{q}_i)$. By linearizing Eq. (6. 29) around the origin and then discretizing it with a sampling time 0. 2 s, we get

$$\boldsymbol{x}_i(k+1) = \begin{bmatrix} 0.9017 & 0.1934 \\ -0.9668 & 0.9013 \end{bmatrix} \boldsymbol{x}_i(k) + \begin{bmatrix} 0.0197 \\ 0.1934 \end{bmatrix} \boldsymbol{u}_i(k) \tag{6.30}$$

Define their neighboring sets of the four manipulators as: $\mathcal{N}_1 = \{4\}$, $\mathcal{N}_2 = \{1\}$, $\mathcal{N}_3 = \{2\}$, $\mathcal{N}_4 = \{3\}$. The design parameters and matrices in *Problem* $\boldsymbol{P}_i$, $\forall i \in \{1,2,3,4\}$ are chosen based on the condition (6. 22) in Theorem 6. 1, and are set as: $\eta = 7$, $\overline{T}_i = 6$, $\gamma_i = 0.13$, $\boldsymbol{Q}_i = 10I$, $\boldsymbol{R}_i = 5\boldsymbol{I}$,

$$\boldsymbol{P}_i = \begin{bmatrix} 16.6152 & 0.05 \\ 0.05 & 1.6582 \end{bmatrix}, \forall i \in \{1,2,3,4\}; \boldsymbol{Q}_{ij} = \begin{bmatrix} 5 & 0 \\ 0 & 3 \end{bmatrix}$$

and $\mu_j = 0.9$ for any $j \in \mathcal{N}_i$. With the initial conditions for $x_1(0) = (1, 1)$, $x_2(0) = (1, -1)$, $x_3(0) = (-1, 1)$, $x_4(0) = (-1, -1)$, the proposed self-triggered DMPC Algorithm 1 is executed and the corresponding simulation results are depicted in Figs. 6.2, 6.3, 6.4 and 6.5. Figs. 6.2 and 6.3 show the angle and angular velocity of the four manipulators respectively, from which it can be seen that the four manipulators finally realizes the desired cooperation task, that is, angle consensus of their single link. The control inputs at each time instant of all the four manipulators are presented in Fig. 6.4 and the triggering instant sequence of each robot manipulator is shown in Fig. 6.5, where vertical coordinate denotes the communication interval optimized at each triggering instant. It is obvious that controllers of the four manipulators are activated in an aperiodic and asynchronous manner rather than a time-triggered periodic way. Therefore, the global control objective is achieved while effectively utilizing communication and computation resources.

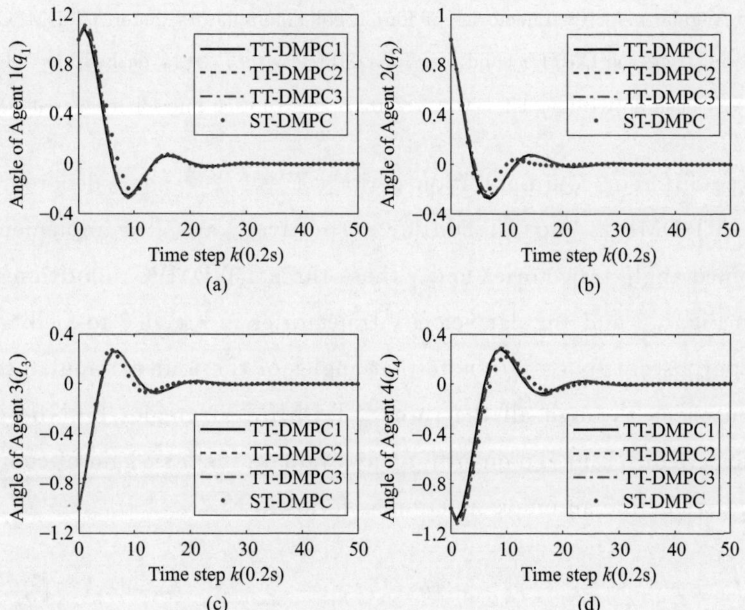

Fig. 6.2 Angle trajectories of four robot manipulators under TT-DMPC and the proposed self-triggered DMPC: solid line-TT-DMPC1 with $T_i = 1$, dashed line-TT-DMPC2 with $T_i = 2$, dashdotted line-TT-DMPC3 with $T_i = 3$, dotted line-self-triggered DMPC

To display the performance level under Algorithm 1, TT-DMPC strategies

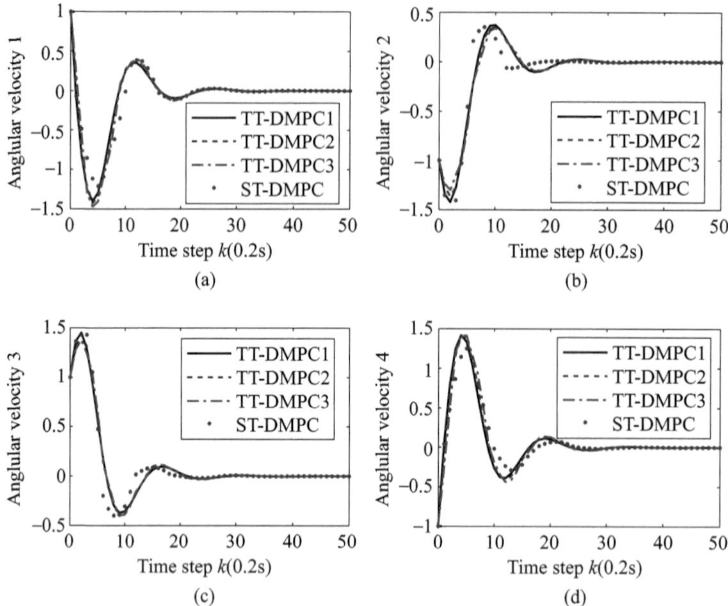

Fig. 6. 3 Angular velocity trajectories of four robot manipulators under TT-DMPC and the proposed self-triggered DMPC: solid line-TT-DMPC1 with $T_i=1$, dashed line-TT-DMPC2 with $T_i=2$, dash-dotted line-TT-DMPC3 with $T_i=3$, dotted line-self-triggered DMPC

with three different communication periods $T=1,2$, and 3, denoted by TT-DMPC1, TT-DMPC2 and TT-DMPC3 respectively, are also implemented. All the obtained angle trajectories under these three TT-DMPC algorithms are depicted in Fig. 6. 2 and angular velocity trajectories in Fig. 6. 3 to facilitate comparison purposes. It is easy to note that angles of the four manipulators finally reach consensus through all the three TT-DMPC strategies but the dynamic properties are different. To quantitatively evaluate the loss/gain of control performance, we define an index function as follows:

$$J_i^c(k)=\sum_{l=0}^{k}\|\boldsymbol{x}_i(l)\|_{\boldsymbol{Q}_i}^2+\|\boldsymbol{u}_i(l)\|_{\boldsymbol{R}_i}^2+\sum_{j\in\mathcal{N}_i}\|\boldsymbol{x}_i(l)-\boldsymbol{x}_j(l)\|_{\boldsymbol{Q}_{ij}}^2 \quad (6.31)$$

where $k=0,\ldots,t_{run}$. The control performance of the overall system is expressed by the sum of index functions of all manipulators, that is,

$$J^c(k)=\sum_{i\in\mathcal{M}}J_i^c(k) \quad (6.32)$$

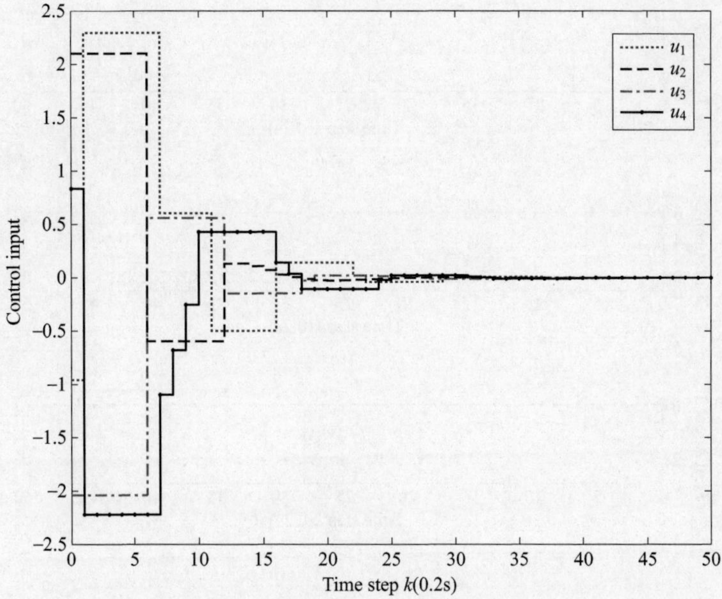

Fig. 6. 4 Control inputs of four robot manipulators under self-triggered DMPC

Then all the control performance trajectories $J^c(k)$ under TT-DMPC1. TT-DMPC2, TT-DMPC 3 and the self-triggered DMPC are illustrated in Fig. 6. 6. By comparing the corresponding $J^c(k)$ under three TT-DMPC strategies, it obtains that the control performance is degraded as communication period increases. Besides, the trajectory of $J^c(k)$ obtained via the proposed self-triggered DMPC is quite near the one obtained by TT-DMPC2. Specifically, its performance index at the steady time is 3. 12 and 0. 89% more than the ones under TT-DMPCI and 2 respectively, which indicates that it is an effective way to adaptively adjust communication period under the proposed self-triggered DMPC in order to obtain a satisfied global performance. Moreover, define the average communication period by $T_{ave,i} = \sum_{h=1}^{d_i} T_i^h / d_i, i = 1,2,3,4$, where d_i is the total triggered number of manipulator i. After computing all the $T_{ave,i}$ under those four control strategies, we realize that average communication periods of the four manipulators are the same under TT-DMPCs, but different from each other under the proposed self-triggered DMPC. More specifically, the average communi-

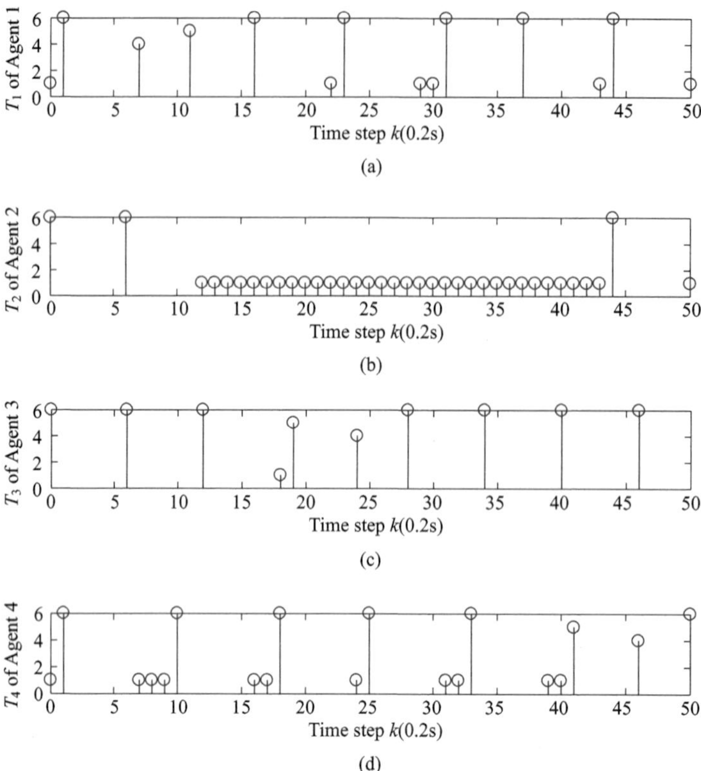

Fig. 6. 5 Triggering instants of four robot manipulators under self-triggered DMPC: from top to bottom the subplots are robot manipulator 1,2,3 and 4 in order

cation periods under the self-triggered DMPC are $T_{ave,1}=3.6429$, $T_{ave,2}=1.4167$, $T_{ave,3}=5.2$, $T_{ave,4}=2.9474$, three of which are close to or longer than 3, thus decreasing communication cost. In Table 6. 1, all communication period $T_{ave,i}$, control performance $J^c(t_{run})$ and $J(t_{run})=\sum_{i\in\mathcal{M}} e^{-\gamma_i T_{ave,i}} J_i^c(t_{run})$ which is the quantitative value of both control performance and communication cost are listed. Since the communication periods of each manipulator under the self-triggered DMPC are different, they are not shown in Table 6. 1. By comparing the proposed self-triggered DMPC with the three TT-DMPCs in $T_{ave,i}$, $J^c(t_{run})$ and $J(t_{run})$, it is concluded that the self-triggered DMPC can achieve a good trade-off between control performance and communication cost.

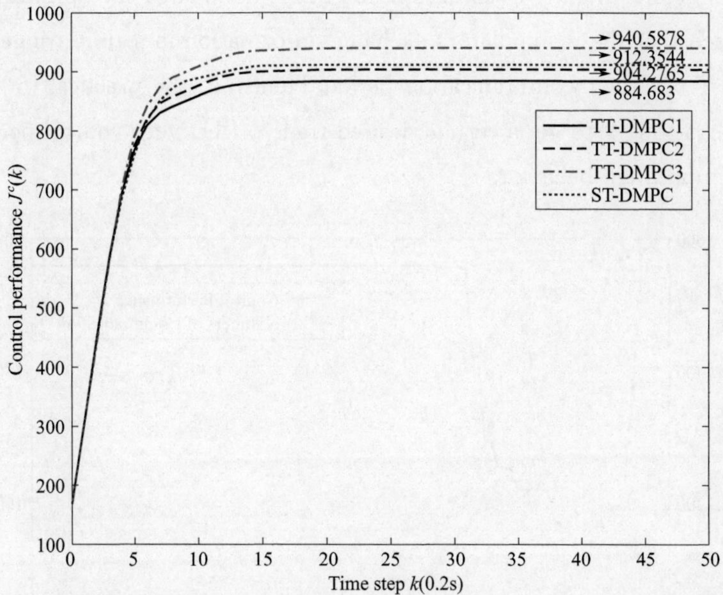

Fig. 6. 6 $J^c(k)$ under TT-DMPC and the proposed self-triggered DMPC: solid line-TT-DMPC1 with $T_i=1$, dashed line-TT-DMPC2 with $T_i=2$, dash-dotted line-TT-DMPC3 with $T_i=3$, dotted line-self-triggered DMPC

Table 6. 1 Control results under TT-DMPC and the proposed self-triggered DMPC

Index	TT-DMPC1	TT-DMPC2	TT-DMPC3	Self-triggered DMPC
$T_{ave,i}$	1	2	3	
$J^c(t_{run})$	884.6837	904.2765	940.5878	912.3544
$J(t_{run})$	776.8367	697.2438	636.8314	581.7567

To illustrate the tuning effect of γ_i on control performance and communication cost, we select about 100 different values of γ from 0 to 1. Let $\gamma_i=\gamma, i=1, 2, 3, 4$, and then carry out the proposed self-triggered DMPC under the same simulation conditions as above. The simulation results are exhibited in Fig. 6. 7, where its vertical coordinate on the right hand represents the sum of communication numbers and is used to reflect the communication cost considering that communication cost is generally proportional to the numbers of communication. It is observed from Fig. 6. 7 that, when γ is less than 0. 1, control performance is mainly concerned at the expense of higher communication cost; as γ goes up, communication cost is down but control performance becomes degraded ; especially when γ is greater than 0.5, communication cost dominates the optimization procedure such that the manipulators exchange information in a time-triggered manner with $\overline{T}_i$ as their communication period. Hence, in the practical implementation, we can tune γ to get a certain desired trade-off between control performance and communication behavior.

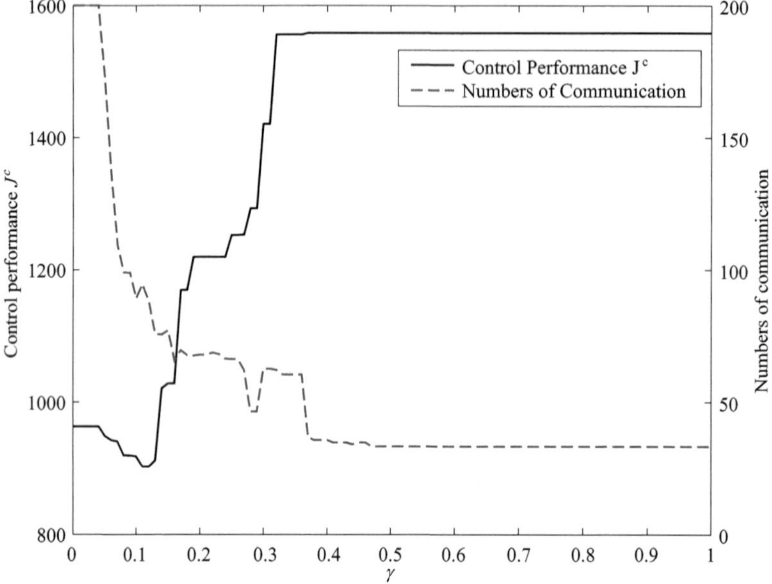

Fig. 6.7 Relation curve of J^c with γ: solid line, relation curve of numbers of communications with γ: dashed line

6.6 Conclusion

In this chapter, the co-design of self-triggered mechanism and DMPC for linear NCSs without physical constraints has been studied and a self-triggered DMPC algorithm has been presented. In this algorithm, control inputs and triggering instants are jointly determined by minimizing the redefined cost function including control performance and communication cost. Then sufficient conditions on the main parameters have been constructed to guarantee stability of the overall system, and simulations on the proposed algorithm have been made and shown its effectiveness. A potential future work would focus on the constrained NCSs, and mainly considers the problem of how to solve the formulated optimization problem effectively and quickly to relieve the on-line computation load and improve real timing.

References

1. Heemels, W. P., Johansson, K. H., & Tabuada, P. (2012). An introduction to event-triggered and self-triggered control. In *2012 IEEE 51th Conference on Decision and Control (CDC)* (pp. 3270-3285). IEEE.
2. Xu. W., Wang. Z., & Ho, D. W. (2017). Finite-horizon H_∞ consensus for multiagent systems with redundant channels via an observer-type event-triggered scheme. *IEEE Transactions on Cybernetics*, *48*(5), 1567-1576.
3. Shen, H., Li, F., Yan, H., Karimi, H. R., & Lam, H. K. (2018). Finite-time event-triggered $\mathcal{H}_\infty$ control for T-S fuzzy markov jump systems. *IEEE Transactions on Fuzzy Systems*, *26*(5), 3122-3135.
4. Zou, Y., Su, X., & Niu, Y. (2016). Event-triggered distributed predictive control for the cooperation of multi-agent systems. *IET Control Theory and Applications*, *11*(1), 10-16.
5. Mi, X., Zou, Y., & Li, S. (2018). Event-triggered MPC design for distributed systems toward global performance. *International Journal of Robust and Nonlinear Control*, *28*(4), 1474-1495.
6. Andrieu, V., Nadri, M., Serres, U., & Vivalda, J. C. (2013). Continuous discrete observer with updated sampling period. *IFAC Proceedings Volumes*, *46*(23), 439-444.
7. Wang, X., & Lemmon, M. D. (2009). Self-triggered feedback control systems with finite-gain $\mathcal{L}_2$ stability. *IEEE Transactions on Automatic Control*, *54*(3), 452-467.
8. Hu, W., Liu, L., & Feng, G. (2016). Output consensus of heterogeneous linear multi-agent systems by distributed event-triggered/self-triggered strategy. *IEEE Transactions on Cybernetics*, *47*(8), 1914-1924.
9. Gommans, T., Antunes, D., Donkers, T., Tabuada, P., & Heemels, M. (2014). Self-triggered linear quadratic control. *Automatica*, *50*(4), 1279-1287.
10. Almeida, J., Silvestre, C., & Pascoal, A. M. (2015). Self-triggered state-feedback control of linear plants under bounded disturbances. *International Journal of Robust and Nonlinear Control*, *25*(8), 1230-1246.
11. Gommans, T. M. P., & Heemels, W. P. M. H. (2015). Resource-aware MPC for constrained nonlinear systems: A self-triggered control approach. *Systems and Control Letters*, *79*, 59-67.
12. Antunes, D., & Heemels, W. P. M. H. (2014). Rollout event-triggered control: Beyond periodic control performance. *IEEE Transactions on Automatic Control*, *59*(12), 3296-3311.
13. Hashimoto, K., Adachi, S., & Dimarogonas, D. V. (2016). Self-triggered model predictive control for continuous-time systems: A multiple discretizations approach. In *2016 IEEE 55th Conference on Decision and Control (CDC)* (pp. 3078-3083). IEEE.

14. Brunner, F. D., Heemels, M., & Allgöwer, F. (2016). Robust self-triggered MPC for constrained linear systems: A tube-based approach. *Automatica*, *72*, 73-83.
15. Henriksson, E., Quevedo, D. E., Peters, E. G., Sandberg, H., & Johansson, K. H. (2015). Multiple-loop self-triggered model predictive control for network scheduling and control. *IEEE Transactions on Control Systems Technology*, *23*(6), 2167-2181.
16. Zhan, J., Li, X., & Jiang, Z. P. (2017). Self-triggered robust output feedback model predictive control of constrained linear systems. In *2017 American Control Conference (ACC)* (pp. 3066-3071). IEEE.
17. Li, H., Yan, W., & Shi, Y. (2017). Adaptive self-triggered model predictive control of discrete-time linear systems. In *2017 IEEE 56th Annual Conference on Decision and Control (CDC)* (pp. 6165-6170). IEEE.
18. Li, H., Yan, W., & Shi, Y. (2018). Triggering and control codesign in self-triggered model predictive control of constrained systems: With guaranteed performance. *IEEE Transactions on Automatic Control*, *63*(11), 4008-4015.
19. Eqtami, A., Heshmati-Alamdari, S., Dimarogonas, D. V., & Kyriakopoulos, K. J. (2013). A self-triggered model predictive control framework for the cooperation of distributed nonholonomic agents. In *2013 IEEE 52nd annual conference on decision and control* (pp. 7384-7389). IEEE.
20. Hashimoto, K., Adachi, S., & Dimarogonas, D. V. (2015). Distributed aperiodic model predictive control for multi-agent systems. *IET Control Theory and Applications*, *9*(1), 10-20.
21. Wang, W., Li, H., Yan, W., & Shi, Y. (2017). Self-triggered distributed model predictive control of nonholonomic systems. In *2017 11th Asian Control Conference (ASCC)* (pp. *280-285*). IEEE.
22. Mi, X., Zou, Y., Li, S., & Karimi, H. R. (*2019*). Self-triggered DMPC design for cooperative multiagent systems. *IEEE Transactions on Industrial Electronics*, *67*(1), 512-520.
23. Xing, L., Wen, C., Guo, F., Liu, Z., & Su, H. (2016). Event-based consensus for linear multiagent systems without continuous communication. *IEEE Transactions on Cybernetics*, *47* (8), 2132-2142.

Chapter 7
Event-Triggered Distributed Model Predictive Control for Interconnected Networked Systems

7.1 Introduction

There are many practical NCSs consisting of several physically interconnected subsystems, such as process control systems [1], power grid systems [2], and networked freeway traffic systems [3]. Different from the NCSs discussed in the aforementioned chapters, subsystems in the interconnected NCSs can directly act on the dynamic behaviors of other subsystems and causes mutual influences, making it more difficult to design DMPC controllers for satisfied global control performance.

Lots of representative achievements in DMPC for these interconnected NCSs have been collected in the classic reviews [1, 4] and published in [5, 7-11]. According to the coupling source, they are divided into state coupling, control input coupling and constraint coupling. The stabilization of nonlinear NCSs with state coupling was studied in [7], and constructed a consistency constraint in the optimized problem to limit the deviation between the predicted state and its assumed state such that the closed-loop stability and recursive feasibility were guaranteed. The mutual influences caused by input coupling were also considered in [8], and were incorporated in the cost function for performance improvement, in addition to the corresponding consistency constraints on control inputs. Furthermore, [9] investigated the DMPC problem for NCSs with state and input coupling constraints and presented a tightened consistency constraint with a monotonically decreasing bound. In [10], the tube

Y. Zou and S. Li, *Distributed Cooperative Model Predictive Control of Networked Systems*, https://doi.org/10.1007/978-981-19-6084-0_7

based method was adopted instead to reject the mutual influences. To make it less conservative, a second MPC controller was introduced and used to optimize the disturbance rejection control action in [10] and a parametric terminal set was applied in [12].

From the point of view of coordination mode, all the above results are in a time triggered way, which results in a heavy computation and communication burden. To save the computation resource, an event-triggered mechanism was introduced to the network between the sensor node and controller node for nonlinear NCSs under MPC in [13,14]. And further, a triggering condition only based on system information of the local subsystem itself was given in [15, 16], which causes the local controller to be motivated as needed to receive the current sampling, update the control law and send its neighbors the updated information so as to reduce communication load. To avoid continuous supervision of the system and checking of the triggering condition, a self-triggered DMPC methodology was presented in [17]. However, the problem of event-triggered DMPC for the interconnected NCSs becomes non-trivial due to the mutual influences and remains further studied.

Therefore, this chapter pays attention to developing an event-triggered DMPC algorithm for the NCSs with dynamically coupled subsystems, providing an effective solution to dealing with mutual influences on recursive feasibility of DMPC in an aperiodic and asynchronous mode. The two main features are as follows. Firstly, certain constraints relevant to the triggering instant are incorporated to ensure the feasibility and global performance. Secondly, two event-triggering conditions are constructed from the analysis of recursive feasibility and system stability with considering the coupling influences, one of which is only related to the deviation between the current measurement and its predicted value of the local subsystem, the other is based on system information received from its neighbors. If and only if either of the above two triggering conditions is violated, the local controller works in the implementation.

The remainder of this chapter is organized as follows. Section 7.2 states the preliminary details and the control problem. In Section 7.3, the constructed triggering conditions is specifically presented. Section 7.4 describes the proposed event-triggered dual-mode DMPC algorithm. In Section 7.5, recursive feasibility and the closed-loop stability are analyzed. Then simulation results under the proposed algorithm are shown in Section 7.6. Finally, Section 7.7 summarizes this chapter.

Notation: The real space is denoted by $\mathbb{R}$, and the collection of all natural numbers is denoted as $\mathbb{N}$. For any vector $\boldsymbol{x} \in \mathbb{R}^n$, $\|\boldsymbol{x}\|$ denotes the Euclidean norm and $\|\boldsymbol{x}\|_{\boldsymbol{P}} = \sqrt{\boldsymbol{x}^{\mathrm{T}}\boldsymbol{P}\boldsymbol{x}}$ stands for the $\boldsymbol{P}$-weighted norm, where $\boldsymbol{P}$ is a matrix with appropriate dimension. The notation $\boldsymbol{P} \succ 0$ means that the matrix $\boldsymbol{P}$ is a positive definite matrix; the maximum eigenvalue of $\boldsymbol{P}$ is denoted by $\overline{\lambda}(\boldsymbol{P})$ and the minimum one is denoted as $\underline{\lambda}(\boldsymbol{P})$. In addition, set $\{1, \ldots, M\} \subseteq \mathbb{N}$ is denoted as $\mathcal{M}$, and $|\mathcal{M}|$ denotes the number of elements in $\mathcal{M}$. A block-diagonal matrix $\boldsymbol{P}$ with blocks $\boldsymbol{P}_i, i \in \mathcal{M}$ is denoted as $\boldsymbol{P} = \mathrm{diag}(P_1, \ldots, P_M)$. If unnecessary, the timestamp is omitted.

7.2 Optimization Problem Formulation

Consider a linear NCS consisting of M subsystems with state coupling, each of which is described by

$$\boldsymbol{x}_i(k+1) = \boldsymbol{A}_{ii}\boldsymbol{x}_i(k) + \boldsymbol{B}_{ii}\boldsymbol{u}_i(k) + \sum_{j \neq i} \boldsymbol{A}_{ij}\boldsymbol{x}_j(k) \tag{7.1}$$

where $i \in \mathcal{M}$, $\boldsymbol{x}_i(k) \in \mathbb{R}^{n_i}$ and $\boldsymbol{u}_i(k) \in \mathbb{R}^{m_i}$ denote the state and control input of subsystem i at time k respectively, $\boldsymbol{A}_{ii} \in \mathbb{R}^{n_i \times n_i}$ and $\boldsymbol{B}_{ii} \in \mathbb{R}^{n_i \times m_i}$ are system matrices, $\boldsymbol{A}_{ij} \in \mathbb{R}^{n_i \times n_j}$ is the coupling matrix which represents the influence of subsystem j on dynamics of subsystem i. Besides, control inputs are constrained by $\boldsymbol{u}_i \in \mathcal{U}_i \subseteq \mathbb{R}^{m_i}$, where $\mathcal{U}_i$ is a compact set containing the origin.

Subsystem $j\,(j \neq i)$ in the following is called an upstream neighbor of subsystem $i \in \mathcal{M}$ if $\boldsymbol{A}_{ij} \neq 0$. Else if $\boldsymbol{A}_{ji} \neq 0$, subsystem j is defined as a downstream neighbor of subsystem i. The sets of all the upstream neighbors and downstream neighbors of subsystem i is denoted by $\mathcal{N}_i^u$ and $\mathcal{N}_i^d$, respectively. Besides, let N_i stand for the set of all neighbors of subsystem i and thus $\mathcal{N}_i = \mathcal{N}_i^u \cup \mathcal{N}_i^d$.

Based on (7.1), the overall system can be expressed by

$$\boldsymbol{x}(k+1) = \boldsymbol{A}\boldsymbol{x}(k) + \boldsymbol{B}\boldsymbol{u}(k) \tag{7.2}$$

where $\boldsymbol{x}=[x_1^{\mathrm{T}},\ldots,x_M^{\mathrm{T}}]^{\mathrm{T}}\in\mathbb{R}^n$, $n=\sum_{i\in\mathcal{M}}n_i$, $\boldsymbol{u}=[u_1^{\mathrm{T}},\ldots,u_M^{\mathrm{T}}]^{\mathrm{T}}\in\mathcal{U}\subseteq\mathbb{R}^m$, $m=\sum_{i\in\mathcal{M}}m_i$, $\mathcal{U}=\mathcal{U}_1\times\cdots\times\mathcal{U}_M$. $\boldsymbol{A}\in\mathbb{R}^{n\times n}$ is a block matrix with its (i,j) block being $\boldsymbol{A}_{ij}$, $\boldsymbol{B}=\mathrm{diag}(B_{11},\ldots,B_{MM})$.

Assumption 7.1 There exists a control law $\boldsymbol{u}_i(k)=\boldsymbol{K}_{ii}\boldsymbol{x}_i(k)$ for each local subsystem $i\in\mathcal{M}$ in (7.1) such that: (i) $\boldsymbol{A}_{d_{ii}}=\boldsymbol{A}_{ii}+\boldsymbol{B}_{ii}\boldsymbol{K}_{ii}$ is Schur. (ii) Let $\boldsymbol{K}=\mathrm{diag}(K_{11},\ldots,K_{MM})$, $\boldsymbol{A}_d=\boldsymbol{A}+\boldsymbol{BK}$ is Schur.

Lemma 7.1 *Suppose that Assumption 7.1 holds. For any subsystem $i\in\mathcal{M}$, (i) Given that $\boldsymbol{Q}_i>0$ and $\boldsymbol{R}_i>0$, there exists a positive definite matrix $\boldsymbol{P}_i$ satisfying $\boldsymbol{A}_{d_{ii}}^{\mathrm{T}}\boldsymbol{P}_i\boldsymbol{A}_{d_{ii}}-\boldsymbol{P}_i\leqslant-(\boldsymbol{Q}_i+\boldsymbol{K}_{ii}^{\mathrm{T}}\boldsymbol{R}_i\boldsymbol{K}_{ii})$. (ii) Let $\hat{\boldsymbol{A}}_d=diag(A_{d_{11}},\ldots,A_{d_{MM}})$, $\boldsymbol{Q}=diag(Q_1,\ldots,Q_M)$, $\boldsymbol{R}=diag(R_1,\ldots,R_M)$, $\boldsymbol{P}=diag(P_1,\ldots,P_M)$, assume that $\boldsymbol{A}_d^{\mathrm{T}}\boldsymbol{P}\boldsymbol{A}_d-\hat{\boldsymbol{A}}_d^{\mathrm{T}}\boldsymbol{P}\hat{\boldsymbol{A}}_d<(\boldsymbol{Q}+\boldsymbol{K}^{\mathrm{T}}\boldsymbol{R}\boldsymbol{K})/2$. There exists a constant $\varepsilon>0$ such that the set $\phi(\varepsilon)\triangleq\{\boldsymbol{x}\in\mathbb{R}^n:\|\boldsymbol{x}\|_{\boldsymbol{P}}\leqslant\varepsilon\}$ is an invariant set and for any $\boldsymbol{x}\in\phi(\varepsilon)$, $\boldsymbol{Kx}\in\mathcal{U}$[8].*

Remark 7.1 Assumption 7.1 and Lemma 7.1 indicate that design of the local controller requires the dynamic model of the global system. The feedback gain $\boldsymbol{K}_{ii}$ and terminal penalty matrix $\boldsymbol{P}_i$ can be obtained by solving an LMI, a sufficient condition for the input constraints in the terminal sets, for more details see [18]. Other methods are presented in [19].

Assumption 7.2 Assume that each subsystem $i\in\mathcal{M}$ ideally broadcasts to and receives from all its neighbors $i\in\mathcal{N}_i$ system information.

This chapter is devoted to developing an event-triggered DMPC algorithm for the system in (7.2) such that each controller can deal with the influences from its neighbors and autonomously coordinate with each other, thereby stabilizing the overall system while relieving computation and communication burden. In Fig. 7.1, the control framework of the proposed event-triggered DMPC algorithm is briefly depicted, where the event triggers act on both the feedback loop of each local subsystem and the communication channels between it and its neighbors. The local MPC controller is activated to take control action and send the corresponding state information to its neighbors if and only if an event happens in subsystem i. The main challenge lies in how to design triggering conditions in presence of dynamic coupling with considering recursive feasibility and system stability.

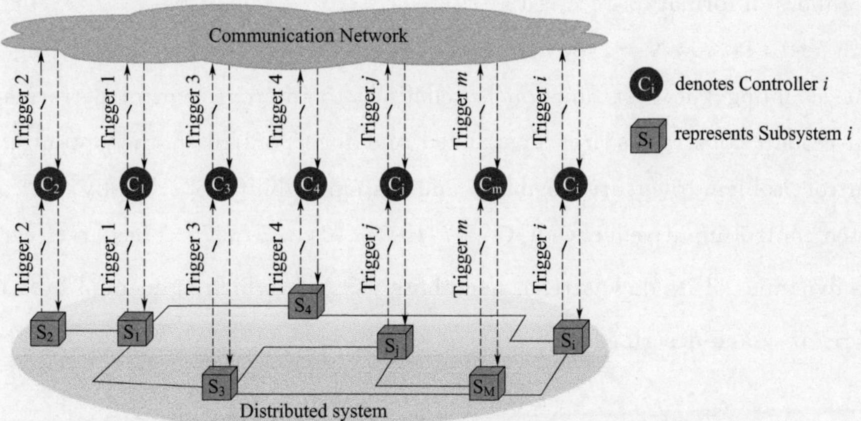

Fig. 7.1 A schematic diagram of the system (2) with the ET-DMPC strategy

The necessary notations are listed here to make it easier for the reader:

$\boldsymbol{x}_i(k)$	The actual state at time k
$\overline{\boldsymbol{u}}_i(k+l\|k)$	The feasible control candidate at time $k+l$ defined at time k.
$\overline{\boldsymbol{x}}_i(k+l\|k)$	The feasible predicted state at time $k+l$ defined at time k, which is obtained based on (7.1) by using $\overline{\boldsymbol{u}}_i(k+l\|k)$ and $\overline{\boldsymbol{x}}_i(k\|k)=\boldsymbol{x}_i(k)$.
$\boldsymbol{u}_i^*(k+l\|k)$	The optimal predicted control at time $k+l$ obtained at time k.
$\boldsymbol{x}_i^*(k+l\|k)$	The optimal predicted state at time $k+l$ obtained at time k, which is obtained based on (7.1) by using $\boldsymbol{u}_i^*(k+l\|k)$ and $\boldsymbol{x}_i^*(k\|k)=\boldsymbol{x}_i(k)$.
$\hat{\boldsymbol{x}}_i(k+N\|k)$	The estimated predicted state at time $k+N$ defined at time k, which is obtained based on the past information.
$\Delta\boldsymbol{x}_{j,i}(k+l\|k)$	The impact of subsystem i on the state of subsystem j at time $k+l$ defined at k.
t_i^r	The r th triggering instant of subsystem i, where $r\in\mathbb{N}$. Let $t_i^0=0$. Note that in general $t_i^r\neq t_j^r$.
$\tau_i(k)$	The largest triggering instant of subsystem i satisfying $\tau_i(k)<k$.

For any subsystem $i\in\mathcal{M}$ at every triggering instant t_i^r, a finite horizon optimization control problem is solved based on the actual state $\boldsymbol{x}_i(t_i^r)$ and its neighbors' information received at $\tau_j(t_i^r)$, i. e. $\boldsymbol{x}_j^*[\tau_j(t_i^r)+l\,|\,\tau_j(t_i^r)], j\in\mathcal{N}_i$ with $l=0,1,\ldots,N-1$, where N is the prediction horizon. Considering the state coupling, the cost function is redefined to improve control performance and certain constraints are constructed and incorporated in the optimization control problem to ensure feasibility and system stability. Specifically, the computed control input sequence $\boldsymbol{u}_i(t_i^r+l\,|\,t_i^r), l=0,1,\ldots,N-1$ has an effect on the dynamics of its downstream neighbors $j\in\mathcal{N}_i^d$, which is denoted by $\Delta\boldsymbol{x}_{j,i}(t_i^r+l\,|\,t_i^r)$ and described according to (7.1) as

$$\Delta\boldsymbol{x}_{j,i}(t_i^r+l\,|\,t_i^r)=\sum_{p=2}^{l}\sum_{h=1}^{p-1}\boldsymbol{A}_{jj}^{l-p}\boldsymbol{A}_{ji}\boldsymbol{A}_{ii}^{p-h-1}\boldsymbol{B}_{ii}[\boldsymbol{u}_i(t_i^r+h-1\,|\,t_i^r)-\boldsymbol{u}_i(t_i^r+h-1\,|\,t_i^r-1)]$$

with $l=2,3,\ldots,N$ and $\Delta\boldsymbol{x}_{j,i}(t_i^r+l\,|\,t_i^r)=\boldsymbol{u}_i(t_i^r-1)-\boldsymbol{u}_i(t_i^r-1\,|\,t_i^r-2)$ for $l=1$, where $\boldsymbol{u}_i(t_i^r+h-1\,|\,t_i^r)$ is the control input to be optimized at t_i^r; $\boldsymbol{u}_i(t_i^r+h-1\,|\,t_i^r-1)$ stands for the predicted control input at t_i^r-1. Note that if t_i^r is just the next sampling instant after t_i^{r-1}, then it holds that $t_i^{r-1}=t_i^r-1$ and $\boldsymbol{u}_i(t_i^r+h-1\,|\,t_i^r-1)=u_i^*(t_i^r+h-1\,|\,t_i^{r-1})$; otherwise, $\boldsymbol{u}_i(t_i^r+h-1\,|\,t_i^r-1)=\bar{\boldsymbol{u}}_i(t_i^r+h-1\,|\,t_i^r-1)$.

Then the control performance of its downstream neighboring subsystem j, denoted by $\tilde{J}_{j,i}\{x_j^*[\tau_j(t_i^r)], u_i(t_i^r)\}$, is approximated by the following function with respect to $\Delta\boldsymbol{x}_{j,i}(t_i^r+l\,|\,t_i^r)$

$$\begin{aligned}&\tilde{J}_{j,i}\{x_j^*[\tau_j(t_i^r)], u_i(t_i^r)\}\\&=\sum_{l=1}^{N-1}\|\boldsymbol{x}_j^*(t_i^r+l\,|\,\tau_j(t_i^r))+\boldsymbol{w}_i\cdot\Delta\boldsymbol{x}_{j,i}(t_i^r+l\,|\,t_i^r)\|_{\boldsymbol{Q}_j}^2\\&+\|\boldsymbol{x}_j^*(t_i^r+N\,|\,\tau_j(t_i^r))+\boldsymbol{w}_i\cdot\Delta\boldsymbol{x}_{j,i}(t_i^r+N\,|\,t_i^r)\|_{\boldsymbol{P}_j}^2\end{aligned}\tag{7.3}$$

where $\boldsymbol{Q}_j\succ 0, \boldsymbol{P}_j\succ 0$, and $\boldsymbol{w}_i\succ 0$ is a weighting factor.

Hence for performance improvement, not only the performance of subsystem i itself but also this approximated performance of its downstream neighbors are involved in the cost function, which is given by

$$
\begin{aligned}
& J_i[\boldsymbol{x}_i(t_i^r), \boldsymbol{u}_i(t_i^r)] \\
& = \sum_{l=0}^{N-1} (\|\boldsymbol{x}_i(t_i^r + l \mid t_i^r)\|_{\boldsymbol{Q}_i}^2 + \|\boldsymbol{u}_i(t_i^r + l \mid t_i^r)\|_{\boldsymbol{R}_i}^2) + \|\boldsymbol{x}_i(t_i^r + N \mid t_i^r)\|_{\boldsymbol{P}_i}^2 \\
& + \sum_{j \in \mathcal{N}_i^d} \mathfrak{J}_{j,i}\{\boldsymbol{x}_j^*[\tau_j(t_i^r)], \boldsymbol{u}_i(t_i^r)\}
\end{aligned} \tag{7.4}
$$

where $\boldsymbol{Q}_i > 0, \boldsymbol{R}_i > 0, \boldsymbol{P}_i > 0$, and $\boldsymbol{P}_i$ is selected based on Lemma 7.1.

Besides, state deviation between the current actual state and its prediction is caused due to the dynamic coupling and the event-triggered communication strategy, which further results in the mismatch between the predicted state and the actual one over the prediction horizon. Thus certain constraints must be constructed to guarantee the feasibility and stability.

To summarize, the optimization control problem for each subsystem $i \in \mathcal{M}$ at its triggering instant t_i^r is formulated: given $\boldsymbol{x}_i(t_i^r), \boldsymbol{x}_i^*(t_i^{r-1} + l \mid t_i^{r-1})$, $\overline{\boldsymbol{x}}_i(t_i^r + l \mid t_i^r), \hat{x}_i(t_i^r + N \mid t_i^r)$ and its neighbors' latest updated information $\boldsymbol{x}_j^*[t_i^r + l \mid \tau_j(t_i^r)], j \in \mathcal{N}_i$ with $l = 0, \ldots, N$, the following Problem 7.1 is solved on-line to obtain $\boldsymbol{U}_i^*(t_i^r) = \{\boldsymbol{u}_i^*(t_i^r \mid t_i^r), \boldsymbol{u}_i^*(t_i^r + 1 \mid t_i^r), \ldots, \boldsymbol{u}_i^*(t_i^r + N - 1 \mid t_i^r)\}$.

Problem 7.1

$$
\min J_i[\boldsymbol{x}_i(t_i^r), \boldsymbol{u}_i(t_i^r)] \tag{7.5}
$$

subject to:

$$
\begin{aligned}
& \boldsymbol{x}_i(t_i^r + l + 1 \mid t_i^r) = \boldsymbol{A}_{ii}\boldsymbol{x}_i(t_i^r + l \mid t_i^r) + \\
& \boldsymbol{B}_{ii}\boldsymbol{u}_i(t_i^r + l \mid t_i^r) + \sum_{j \in \mathcal{N}_i^u} \boldsymbol{A}_{ij}\boldsymbol{x}_j^*[t_i^r + l \mid \tau_j(t_i^r)]
\end{aligned} \tag{7.6}
$$

$$
\begin{aligned}
& \sum_{h=0}^{l-1} \pi(l - h - 1) \|\boldsymbol{x}_i(t_i^r + h \mid t_i^r) - \boldsymbol{x}_i^*(t_i^r + h \mid t_i^{r-1})\| \\
& \leqslant \frac{\gamma\kappa\varepsilon}{2Mm_1}, l = 1, \ldots, N - 1
\end{aligned} \tag{7.7}
$$

$$
\|\boldsymbol{x}_i(t_i^r + N \mid t_i^r) - \hat{\boldsymbol{x}}_i(t_i^r + N \mid t_i^r)\|_{\boldsymbol{P}_i} \leqslant \frac{\kappa\varepsilon}{2M} \tag{7.8}
$$

$$
\|\boldsymbol{x}_i(t_i^r + l \mid t_i^r)\|_{\boldsymbol{P}_i} \leqslant \|\overline{\boldsymbol{x}}_i(t_i^r + l \mid t_i^r)\|_{\boldsymbol{P}_i} + \frac{\varepsilon}{\rho MN}, l = 1, \ldots, N \tag{7.9}
$$

$$\boldsymbol{x}_i(t_i^r+N|t_i^r)\in\phi_i\left(\frac{\varepsilon}{2M}\right) \tag{7.10}$$

$$\boldsymbol{u}_i(t_i^r+l|t_i^r)\in\boldsymbol{\mathcal{U}}_i, l=0,\ldots,N-1 \tag{7.11}$$

where $\boldsymbol{x}_i(t_i^r|t_i^r)=\boldsymbol{x}_i(t_i^r)$, $\pi(s)=\max_{i,j\in\mathcal{M},i\neq j}\lambda^{-\frac{1}{2}}[(\boldsymbol{A}_{ii}^s\boldsymbol{A}_{ij})^{\mathrm{T}}\boldsymbol{P}_i\boldsymbol{A}_{ii}^s\boldsymbol{A}_{ij}]$, $m_1=\max_{i\in\mathcal{M}}|\mathcal{N}_i^u|$, $\tau_j(t_i^r)$ is the largest triggering instant t_j^g of subsystem j satisfying $t_j^g<t_i^r$, $\phi_i\left(\frac{\varepsilon}{2M}\right)\triangleq\left\{\boldsymbol{x}_i\in\mathbb{R}^{n_i}:\|\boldsymbol{x}_i\|_{\boldsymbol{P}_i}\leqslant\frac{\varepsilon}{2M}\right\}$ is the terminal set, γ and κ are in $(0,1)$ and $\rho>2$, whose selection is further discussed in Remarks 7.4 and 7.5 respectively. $\hat{\boldsymbol{x}}_i(t_i^r+N|t_i^r)$ is denoted as the estimated predicted state at t_i^r+N calculated at t_i^r, in the following its general definition at every sampling instant k is given in detail.

(i) For $k=t_i^{r-1}+1$, which indicates that controller i was triggered at the last sampling instant, $\boldsymbol{x}_i^*(t_i^{r-1}+N|t_i^{r-1})$ was known, then $\hat{\boldsymbol{x}}_i(k+N|k)$ is defined by

$$\hat{\boldsymbol{x}}_i(k+N|k)=\boldsymbol{A}_{d_{ii}}\boldsymbol{x}_i^*(t_i^{r-1}+N|t_i^{r-1})+\sum_{j\in\mathcal{N}_i^u}\boldsymbol{A}_{ij}\boldsymbol{x}_j^*[k+N-1|\tau_j(k)] \tag{7.12}$$

(ii) For $t_i^{r-1}+1<k\leqslant t_i^r$, the constructed input candidate $\overline{u}_i(k+l-1|k-1)$ were used and $\overline{\boldsymbol{x}}_i(k+l|k-1), l=0,\ldots,N-1$ were obtained at the last sampling instant, thus $\hat{\boldsymbol{x}}_i(k+N|k)$ is defined by

$$\hat{\boldsymbol{x}}_i(k+N|k)=\boldsymbol{A}_{d_{ii}}\overline{\boldsymbol{x}}_i(k+N-1|k-1)+\sum_{j\in\mathcal{N}_i^u}\boldsymbol{A}_{ij}\boldsymbol{x}_j^*[k+N-1|\tau_j(k)] \tag{7.13}$$

Besides, the state trajectory $\overline{\boldsymbol{x}}_i$ in (7.9) and (7.13) is calculated by using the constructed feasible input candidate $\overline{u}_i$ based on (7.1), which is specifically described as follows:

(i) For $k=t_i^{r-1}+1$, the feasible input candidate $\overline{\boldsymbol{u}}_i(k+l|k)$ based on the optimal control input at t_i^{r-1} is constructed by

$$\overline{\boldsymbol{u}}_i(k+l|k)=\begin{cases}\boldsymbol{u}_i^*(k+l|t_i^{r-1}), l=0,\ldots,N-2\\ \boldsymbol{K}_{ii}\overline{\boldsymbol{x}}_i(k+l|k), l=N-1\end{cases} \tag{7.14}$$

Let $\overline{\boldsymbol{x}}_i(k|k)=\boldsymbol{x}_i(k)$, the corresponding predicted state $\overline{\boldsymbol{x}}_i(k+l|k)$ with $l=1,2,\ldots,N$ can be obtained by:

$$\overline{\boldsymbol{x}}_i(k+l+1|k)=\boldsymbol{A}_{ii}\overline{\boldsymbol{x}}_i(k+l|k)+\boldsymbol{B}_{ii}\overline{\boldsymbol{u}}_i(k+l|k)+\sum_{j\in\mathcal{N}_i^u}\boldsymbol{A}_{ij}\boldsymbol{x}_j^*[k+l|\tau_j(k)] \tag{7.15}$$

(ii) For $t_i^{r-1}+1<k\leqslant t_i^r$, controller i is not activated to solve Problem 7.1 at $k-1$, and generates a feasible input candidate sequence $\overline{\boldsymbol{u}}_i(k+l|k-1)$ based on $\overline{\boldsymbol{u}}_i(k+l|k-1)$

$$\overline{\boldsymbol{u}}_i(k+l|k)=\begin{cases}\overline{\boldsymbol{u}}_i(k+l|k-1), l=0,\ldots,N-2\\ \boldsymbol{K}_{ii}\overline{\boldsymbol{x}}_i(k+l|k), l=N-1\end{cases} \tag{7.16}$$

Let $\overline{\boldsymbol{x}}_i(k|k)=\boldsymbol{x}_i(k)$, the corresponding predicted state $\overline{\boldsymbol{x}}_i(k+l|k)$ with $l=1,2,\ldots,N$ can be obtained according to (7.15) and (7.16).

Remark 7.2 In the proposed event-triggered DMPC algorithm, subsystem i will not send its state information to its neighbors at any instant $k\in(t_i^{r-1}, t_i^r)$. In this case, $\boldsymbol{x}_i^*(k+l|t_i^{r-1}), l=0,1,\ldots,t_i^{r-1}+N-k$ of being broadcast at t_i^{r-1} will be used as the assumed state trajectory by its neighbors, which deviates from $\boldsymbol{x}_i(k+l|t_i^r)$ the one generated by local subsystem i. Therefore, constraint (7.7) is imposed to confine $\boldsymbol{x}_i(k+l|t_i^r)$ close to $\boldsymbol{x}_i^*(k+l|t_i^{r-1})$, which are extremely important to bound the mutual influences and ensure that the constructed input sequence $\overline{\boldsymbol{u}}_i(t_i^r+l|t_i^r)$ is a feasible solution to Problem 7.1 at t_i^r along with constraint (7.8). Besides, constraint (7.9) is constructed to ensure system stability, and a tightened terminal constraint (7.10) is adopted to drive the overall system to the terminal set $\phi(\varepsilon)$. The theoretical analysis of feasibility and stability will be represented in Section 7.5.

7.3 Event-Triggering Conditions for NCSs with Dynamic Coupling

In this section, the event-triggering conditions will be constructed with considering the mutual influences on system properties and reducing communication cost.

Without loss of generality, the following is discussed for any two successive triggering instants t_i^r and t_i^{r+1} of any subsystem $i \in \mathcal{M}$. Firstly, denote the optimal control input sequence of Problem 7.1 at t_i^r and the corresponding optimal state trajectory as $\boldsymbol{u}_i^*(t_i^r+l\,|\,t_i^r)$ and $\boldsymbol{x}_i^*(t_i^r+l\,|\,t_i^r)$ respectively, the feasible control sequence at time instant $k \in (t_i^r, t_i^{r+1})$ is defined by $\overline{\boldsymbol{u}}_i(t_i^r+l\,|\,t_i^r)$. Then a local Lyapunov function is defined for subsystem i as

$$\mathbf{V}_i[\boldsymbol{x}_i(k\,|\,k)] = \sum_{l=0}^{N} \|\boldsymbol{x}_i(k+l\,|\,k)\|_{\boldsymbol{P}_i} \tag{7.17}$$

A triggering condition will be constructed to guarantee that the above defined Lyapunov function is monotonically decreasing, which cannot raise the communication network connectivity. In the following, the difference between $\mathbf{V}_i[\boldsymbol{x}_i(k\,|\,k)]$ and $\mathbf{V}_i[\boldsymbol{x}_i(k+1\,|\,k+1)]$ will be analyzed, where two cases are discussed: (i) For $k=t_i^r+1$, which means that an event just happens at $k-1=t_i^r$, thus $\mathbf{V}_i[\overline{\boldsymbol{x}}_i(k\,|\,k)]-\mathbf{V}_i[\boldsymbol{x}_i^*(t_i^r\,|\,t_i^r)]$ will be studied; (ii) For $t_i^r+1<k<t_i^{r+1}$, that is, the controller is not activated at the two instants $k-1$ and k, hence $\mathbf{V}_i[\overline{\boldsymbol{x}}_i(k\,|\,k)]-\mathbf{V}_i[\overline{\boldsymbol{x}}_i(k-1\,|\,k-1)]$ will be investigated.

(i) For $k=t_i^r+1$, $\mathbf{V}_i[\boldsymbol{x}_i^*(t_i^r\,|\,t_i^r)]$ is obtained by (7.17) based on $\boldsymbol{x}_i^*(t_i^r+l\,|\,t_i^r)$, and then the difference between $\mathbf{V}_i[\overline{\boldsymbol{x}}_i(k\,|\,k)]$ and $\mathbf{V}_i[\boldsymbol{x}_i^*(t_i^r\,|\,t_i^r)]$ is calculated by:

$$\begin{aligned}
&\mathbf{V}_i[\overline{\boldsymbol{x}}_i(k\,|\,k)]-\mathbf{V}_i[\boldsymbol{x}_i^*(t_i^r\,|\,t_i^r)]\\
&=\sum_{l=0}^{N}[\|\overline{\boldsymbol{x}}_i(k+l\,|\,k)\|_{\boldsymbol{P}_i}-\|\boldsymbol{x}_i^*(t_i^r+l\,|\,t_i^r)\|_{\boldsymbol{P}_i}]\\
&=\sum_{l=1}^{N=1}[\|\overline{\boldsymbol{x}}_i(k+l\,|\,k)\|_{\boldsymbol{P}_i}-\|\boldsymbol{x}_i^*(k+l\,|\,t_i^r)\|_{\boldsymbol{P}_i}]+\|\overline{\boldsymbol{x}}_i(k+N\,|\,k)\|_{\boldsymbol{P}_i}\\
&+\|\overline{\boldsymbol{x}}_i(k\,|\,k)\|_{\boldsymbol{P}_i}-\|\boldsymbol{x}_i^*(k\,|\,t_i^r)\|_{\boldsymbol{P}_i}-\|\boldsymbol{x}_i^*(t_i^r\,|\,t_i^r)\|_{\boldsymbol{P}_i}\leqslant\|\boldsymbol{x}_i(k)-\boldsymbol{x}_i^*(k\,|\,t_i^r)\|_{\boldsymbol{P}_i}\\
&+\sum_{l=1}^{N-1}\|\overline{\boldsymbol{x}}_i(k+l\,|\,k)-\boldsymbol{x}_i^*(k+l\,|\,t_i^r)\|_{\boldsymbol{P}_i}\\
&+\|\overline{\boldsymbol{x}}_i(k+N\,|\,k)-\hat{\boldsymbol{x}}_i(k+N\,|\,k)\|_{\boldsymbol{P}_i}+\|\hat{\boldsymbol{x}}_i(k+N\,|\,k)\|_{\boldsymbol{P}_i}-\|\boldsymbol{x}_i(t_i^r)\|_{\boldsymbol{P}_i}
\end{aligned} \tag{7.18}$$

According to (7.15), it is derived for $l=1,\dots,N-1$ that

$$\overline{\boldsymbol{x}}_i(k+l\,|\,k)=\boldsymbol{A}_{ii}^l\boldsymbol{x}_i(k)+\sum_{h=0}^{l-1}\boldsymbol{A}_{ii}^{l-h-1}\boldsymbol{B}_{ii}\overline{\boldsymbol{u}}_i(k+h\,|\,k)$$

$$+\sum_{j\in\mathcal{N}_i^u}\sum_{h=0}^{l-1}\boldsymbol{A}_{ii}^{l-h-1}\boldsymbol{A}_{ij}\boldsymbol{x}_j^*[k+h\,|\,\tau_j(k)] \tag{7.19}$$

Similarly, $\boldsymbol{x}_i^*(k+l\,|\,t_i^r)$ is expressed by

$$\boldsymbol{x}_i^*(k+l\,|\,t_i^r)=\boldsymbol{A}_{ii}^l\boldsymbol{x}_i^*(k\,|\,t_i^r)+\sum_{h=0}^{l-1}\boldsymbol{A}_{ii}^{l-h-1}\boldsymbol{B}_{ii}\boldsymbol{u}_i^*(k+h\,|\,t_i^r)$$
$$+\sum_{j\in\mathcal{N}_i^u}\sum_{h=0}^{l-1}\boldsymbol{A}_{ii}^{l-h-1}\boldsymbol{A}_{ij}\boldsymbol{x}_j^*[k+h\,|\,\tau_j(t_i^r)] \tag{7.20}$$

Since $\overline{\boldsymbol{u}}_i(k+h\,|\,k)=\boldsymbol{u}_i^*(k+h\,|\,t_i^r)$ holds for any $h=0,\ldots,N-2$, by subtracting (7.20) from (7.19), we get

$$\begin{aligned}&\overline{\boldsymbol{x}}_i(k+l\,|\,k)-\boldsymbol{x}_i^*(k+l\,|\,t_i^r)\\&=\boldsymbol{A}_{ii}^l[\boldsymbol{x}_i(k)-\boldsymbol{x}_i^*(k\,|\,t_i^r)]+\sum_{j\in\mathcal{N}_i^u}\sum_{h=0}^{l-1}\boldsymbol{A}_{ii}^{l-h-1}\boldsymbol{A}_{ij}\{\boldsymbol{x}_j^*[k+h\,|\,\tau_j(k)]\\&-\boldsymbol{x}_j^*[k+h\,|\,\tau_j(t_i^r)]\}\\&=\boldsymbol{A}_{ii}^l\boldsymbol{e}_i(k)+\sum_{j\in\mathcal{N}_i^u}\Gamma_j\{\eta_j[k+l\,|\,\tau_j(k)]\}\end{aligned} \tag{7.21}$$

where $\boldsymbol{e}_i(k)=\boldsymbol{x}_i(k)-\boldsymbol{x}_i^*(k\,|\,t_i^r)$, $\eta_j[k+h\,|\,\tau_j(k)]=\boldsymbol{x}_j^*[k+h\,|\,\tau_j(k)]-\boldsymbol{x}_j^*[k+h\,|\,\tau_j(t_i^r)]$, and $\Gamma_j\{\eta_j[k+l\,|\,\tau_j(k)]\}=\sum_{h=0}^{l-1}\boldsymbol{A}_{ii}^{l-h-1}\boldsymbol{A}_{ij}\eta_j[k+h\,|\,\tau_j(k)]$

For $l=N$, according to (7.14) and (7.15), there is

$$\begin{aligned}&\overline{\boldsymbol{x}}_i(k+N\,|\,k)\\&=\boldsymbol{A}_{ii}\overline{\boldsymbol{x}}_i(k+N-1\,|\,k)+\boldsymbol{B}_{ii}\boldsymbol{K}_{ii}\overline{\boldsymbol{x}}_i(k+N-1\,|\,k)\\&\quad+\sum_{j\in\mathcal{N}_i^u}\boldsymbol{A}_{ij}\boldsymbol{x}_j^*[k+N-1\,|\,\tau_j(k)]\\&=\boldsymbol{A}_{d_{ii}}\overline{\boldsymbol{x}}_i(k+N-1\,|\,k)+\sum_{j\in\mathcal{N}_i^u}\boldsymbol{A}_{ij}\boldsymbol{x}_j^*[k+N-1\,|\,\tau_j(k)]\end{aligned} \tag{7.22}$$

Subtracting (7.12) from (7.22) and then using (7.21), we have

$$\begin{aligned}&\overline{\boldsymbol{x}}_i(k+N\,|\,k)-\hat{\boldsymbol{x}}_i(k+N\,|\,k)\\&=\boldsymbol{A}_{d_{ii}}[\overline{\boldsymbol{x}}_i(k+N-1\,|\,k)-\boldsymbol{x}_i^*(k+N-1\,|\,t_i^r)]\\&=\boldsymbol{A}_{d_{ii}}\boldsymbol{A}_{ii}^{N-1}e_i(k)+\sum_{j\in\mathcal{N}_i^u}\boldsymbol{A}_{d_{ii}}\Gamma_j\{\eta_j[k+N-1\,|\,\tau_j(k)]\}\end{aligned} \tag{7.23}$$

Substituting (7.21) and (7.23) to (7.18), we further get

$$\begin{aligned}
&\mathbf{V}_i[\bar{\boldsymbol{x}}_i(k|k)]-\mathbf{V}_i[\boldsymbol{x}_i^*(t_i^r|t_i^r)] \\
&\leqslant \| \mathbf{A}_{d_{ii}}\mathbf{A}_{ii}^{N-1}\boldsymbol{e}_i(k)+\sum_{j\in\mathcal{N}_i^u}\mathbf{A}_{d_{ii}}\Gamma_j\{\eta_j[k+N-1|\tau_j(k)]\}\|_{\mathbf{P}_i}+ \\
&\|\hat{\boldsymbol{x}}_i(k+N|k)\|_{\mathbf{P}_i}+\sum_{l=1}^{N-1}\|\mathbf{A}_{ii}^l\boldsymbol{e}_i(k)+\sum_{j\in\mathcal{N}_i^u}\Gamma_j\{\eta_j[k+l|\tau_j(k)]\}\|_{\mathbf{P}_i} \\
&+\|\boldsymbol{e}_i(k)\|_{\mathbf{P}_i}-\|\boldsymbol{x}_i(t_i^r)\|_{\mathbf{P}_i}
\end{aligned} \tag{7.24}$$

It is found from (7.24) that if

$$\begin{aligned}
&\|\mathbf{A}_{d_{ii}}\mathbf{A}_{ii}^{N-1}\boldsymbol{e}_i(k)+\sum_{j\in\mathcal{N}_i^u}\mathbf{A}_{d_{ii}}\Gamma_j\{\eta_j[k+N-1|\tau_j(k)]\}\|_{\mathbf{P}_i}+\|\hat{\boldsymbol{x}}_i(k+N|k)\|_{\mathbf{P}_i} \\
&+\sum_{l=1}^{N-1}\|\mathbf{A}_{ii}^l\boldsymbol{e}_i(k)+\sum_{j\in\mathcal{N}_i^u}\Gamma_j\{\eta_j[k+l|\tau_j(k)]\}\|_{\mathbf{P}_i}+\|\boldsymbol{e}_i(k)\|_{\mathbf{P}_i}\leqslant\sigma_i\|\boldsymbol{x}_i(t_i^r)\|_{\mathbf{P}_i}
\end{aligned} \tag{7.25}$$

where $0<\sigma_i<1$, it has

$$\mathbf{V}_i[\bar{\boldsymbol{x}}_i(k|k)]-\mathbf{V}_i[\boldsymbol{x}_i^*(t_i^r|t_i^r)]\leqslant(\sigma_i-1)\|\boldsymbol{x}_i(t_i^r)\|_{\mathbf{P}_i}<0$$

To sum up, if condition (7.25) holds, the feasible control input $\bar{\boldsymbol{u}}_i(k|k)$ is able to drive subsystem i convergent from the time instant k to instant $k-1$.

(ii) For $t_i^r+1<k<t_i^{r+1}$, the feasible control inputs $\bar{\boldsymbol{u}}_i(k+l|k-1)$ and $\bar{\boldsymbol{u}}_i(k+l|k)$ are applied at instants $k-1$ and k respectively, and further the corresponding feasible state $\bar{\boldsymbol{x}}_i(k+l|k-1)$ and $\bar{\boldsymbol{x}}_i(k+l|k)$ are used to calculate $\mathbf{V}_i[\bar{\boldsymbol{x}}_i(k-1|k-1)]$ and $\mathbf{V}_i[\bar{\boldsymbol{x}}_i(k|k)]$ separately. Hence, the difference between $\mathbf{V}_i[\bar{\boldsymbol{x}}_i(k|k)]$ and $\mathbf{V}_i[\bar{\boldsymbol{x}}_i(k-1|k-1)]$ is:

$$\begin{aligned}
&\mathbf{V}_i[\bar{\boldsymbol{x}}_i(k|k)]-\mathbf{V}_i[\bar{\boldsymbol{x}}_i(k-1|k-1)] \\
&\leqslant\sum_{l=1}^{N-1}\|\bar{\boldsymbol{x}}_i(k+l|k)-\bar{\boldsymbol{x}}_i(k+l|k-1)\|_{\mathbf{P}_i}+\|\bar{\boldsymbol{x}}_i(k+N|k)-\hat{\boldsymbol{x}}_i(k+N|k)\|_{\mathbf{P}_i} \\
&+\|\hat{\boldsymbol{x}}_i(k+N|k)\|_{\mathbf{P}_i}+\|\boldsymbol{x}_i(k)-\bar{\boldsymbol{x}}_i(k|k-1)\|_{\mathbf{P}_i}-\|\boldsymbol{x}_i(k-1)\|_{\mathbf{P}_i}
\end{aligned} \tag{7.26}$$

According to (7.15) and (7.16), there is for $l=1,\dots,N-1$,

$$\bar{\boldsymbol{x}}_i(k+l|k)-\bar{\boldsymbol{x}}_i(k+l|k-1)=\mathbf{A}_{ii}^l\boldsymbol{e}_i(k)+\sum_{j\in\mathcal{N}_i^u}\Gamma_j\{\eta_j[k+l|\tau_j(k)]\} \tag{7.27}$$

Since $\overline{\boldsymbol{u}}_i(k+l|k-1)$ and $\overline{\boldsymbol{x}}_i(k+l|k-1)$ are used at $k-1$, it holds that $\boldsymbol{e}_i(k)=\boldsymbol{x}_i(k)-\overline{\boldsymbol{x}}_i(k|k-1)$, $\eta_j[k+h|\tau_j(k)]=\boldsymbol{x}_j^*[k+h|\tau_j(k)]-\boldsymbol{x}_j^*[k+h|\tau_j(k-1)]$ in (7.27).

For $l=N$, based on (7.13) and (7.15), and then using (7.27), we obtain that

$$\begin{aligned}&\overline{\boldsymbol{x}}_i(k+N|k)-\hat{\boldsymbol{x}}_i(k+N|k)\\&=\boldsymbol{A}_{d_{ii}}[\overline{\boldsymbol{x}}_i(k+N-1|k)-\overline{\boldsymbol{x}}_i(k+N-1|k-1)]\\&=\boldsymbol{A}_{d_{ii}}\boldsymbol{A}_{ii}^{N-1}\boldsymbol{e}_i(k)+\sum_{j\in\mathcal{N}_i^u}\boldsymbol{A}_{d_{ii}}\Gamma_j\{\eta_j[k+N-1|\tau_j(k)]\}\end{aligned}\tag{7.28}$$

Substituting (7.27) and (7.28) to (7.26), we further get

$$\begin{aligned}&\boldsymbol{V}_i[\overline{\boldsymbol{x}}_i(k|k)]-\boldsymbol{V}_i[\overline{\boldsymbol{x}}_i(k-1|k-1)]\\&\leqslant\|\boldsymbol{A}_{d_{ii}}\boldsymbol{A}_{ii}^{N-1}\boldsymbol{e}_i(k)+\sum_{j\in\mathcal{N}_i^u}\boldsymbol{A}_{d_{ii}}\Gamma_j\{\eta_j[k+N-1|\tau_j(k)]\}\|_{\boldsymbol{P}_i}+\|\hat{\boldsymbol{x}}_i(k+N|k)\|_{\boldsymbol{P}_i}\\&+\sum_{l=1}^{N-1}\|\boldsymbol{A}_{ii}^l\boldsymbol{e}_i(k)+\sum_{j\in\mathcal{N}_i^u}\Gamma_j\{\eta_j[k+l|\tau_j(k)]\}\|_{\boldsymbol{P}_i}+\|\boldsymbol{e}_i(k)\|_{\boldsymbol{P}_i}-\|\boldsymbol{x}_i(k-1)\|_{\boldsymbol{P}_i}\end{aligned}\tag{7.29}$$

Hence, if the following condition holds,

$$\begin{aligned}&\|\boldsymbol{A}_{d_{ii}}\boldsymbol{A}_{ii}^{N-1}\boldsymbol{e}_i(k)+\sum_{j\in\mathcal{N}_i^u}\boldsymbol{A}_{d_{ii}}\Gamma_j\{\eta_j[k+N-1|\tau_j(k)]\}\|_{\boldsymbol{P}_i}+\|\hat{\boldsymbol{x}}_i(k+N|k)\|\boldsymbol{P}_i\\&+\sum_{l=1}^{N-1}\|\boldsymbol{A}_{ii}^l\boldsymbol{e}_i(k)+\sum_{j\in\mathcal{N}_i^u}\Gamma_j\{\eta_j[k+l|\tau_j(k)]\}\|_{\boldsymbol{P}_i}+\|\boldsymbol{e}_i(k)\|_{\boldsymbol{P}_i}\leqslant\sigma_i\|x_i(k-1)\|_{\boldsymbol{P}_i}\end{aligned}\tag{7.30}$$

where $0<\sigma_i<1$, it has

$$\boldsymbol{V}_i[\overline{\boldsymbol{x}}_i(k|k)]-\boldsymbol{V}_i[\overline{\boldsymbol{x}}_i(k-1|k-1)]\leqslant(\sigma_i-1)\|\boldsymbol{x}_i(k-1)\|_{\boldsymbol{P}_i}<0$$

which means that the feasible control input $\overline{\boldsymbol{u}}_i(k|k)$ is capable to preserve the convergence of subsystem i.

Summarizing cases (i) and (ii), the event-triggering condition at instant k is built as follows in order to ensure the convergence of subsystem i:

$$\begin{aligned}&\|\boldsymbol{A}_{d_{ii}}\boldsymbol{A}_{ii}^{N-1}\boldsymbol{e}_i(k)+\sum_{j\in\mathcal{N}_i^u}\boldsymbol{A}_{d_{ii}}\Gamma_j\{\eta_j[k+N-1|\tau_j(k)]\}\|_{\boldsymbol{P}_i}+\|\hat{\boldsymbol{x}}_i(k+N|k)\|_{\boldsymbol{P}_i}\\&+\sum_{l=1}^{N-1}\|\boldsymbol{A}_{ii}^l\boldsymbol{e}_i(k)+\sum_{j\in\mathcal{N}_i^u}\Gamma_j\{\eta_j[k+l|\tau_j(k)]\}\|_{\boldsymbol{P}_i}+\|\boldsymbol{e}_i(k)\|_{\boldsymbol{P}_i}\leqslant\sigma_i\|\boldsymbol{x}_i(k-1)\|_{\boldsymbol{P}_i}\end{aligned}\tag{7.31}$$

where $0<\sigma_i<1$. Note that if an event happens at $k-1$, $\boldsymbol{e}_i(k)=\boldsymbol{x}_i(k)-\boldsymbol{x}_i^*(k|k-1)$; otherwise, $\boldsymbol{e}_i(k)=\boldsymbol{x}_i(k)-\overline{\boldsymbol{x}}_i(k|k-1)$.

Due to the state coupling and the event-triggered coordinated strategy, a state deviation between the current state of subsystem i and its predicted one occurs. which causes the state predicted over the prediction horizon far away from the actual one, and even leads to constraint (7.7) and the terminal constraint (7.10) that cannot be met if the state deviation is too large. Therefore, the other event-triggering condition is developed for subsystem i at k to ensure the recursive feasibility of the Problem 7.1, which is defined by

$$
\begin{aligned}
&\|\boldsymbol{x}_i(k)-\boldsymbol{x}_i^*(k|k-1)\|_{P_i}\leqslant\delta_i,\ \text{if controller } i \text{ is triggered at } k-1\\
&\|\boldsymbol{x}_i(k)-\overline{\boldsymbol{x}}_i(k|k-1)\|_{P_i}\leqslant\delta_i,\ \text{otherwise}
\end{aligned}
\tag{7.32}
$$

Besides, during the DMPC implementation, if neither the constructed triggering condition (7.31) nor (7.32) is violated from instant t_i^r+1 to t_i^r+N, Problem 7.1 is going to be solved at t_i^r+N. In conclusion, the next triggering instant t_i^{r+1} is determined if one of the following event-triggering conditions is violated:

$$
(7.37)\ or\ (7.38)\ or\ k<t_i^r+N \tag{7.33}
$$

7.4 Event-Triggered DMPC Algorithm for Interconnected NCSs

Assume that Problem 7.1 is feasible at the initial instant. To further reduce the consumption of computation and communication resources, an event-triggered DMPC with a dual-mode strategy is developed and is further summarized in Algorithm 1.

Remark 7.3 At the initial instant $k=0$, Problem 7.1 without (7.7)-(7.9) is solved iteratively to obtain the initial control input and the neighbors' state trajectory, referring to Step 3, which can also be gotten by solving the corresponding centralized optimization control problem. θ_i stands for the switching signal. To be more specific, when the system state is outside the defined terminal set, $\theta_i = 0$ and the event-triggered DMPC strategy is applied, see Steps 12-17. Otherwise when the system state converges to this terminal set, θ_i becomes 1 and the control strategy is switched to the feedback control mode, see Step 20. As for Steps 5-10, an iterative algorithm in [19] is employed to check whether the overall system converges to the terminal set or not. Obviously, it is required frequent communication to decide when to switch to the control mode, which can be avoided by making the terminal set sufficiently small so that the event-triggered DMPC strategy is always adopted. It will be proved that the system state is gradually driven to the terminal set in the following section.

Algorithm 1 Event-triggered DMPC for Interconnected NCSs

Off-line: For each controller i, calculate $\boldsymbol{K}_i$ and $\boldsymbol{P}_i$ with the given matrixes $\boldsymbol{Q}_i$, $\boldsymbol{R}_i$ according to Lemma 7.1; choose the parameters κ, γ, ρ and δ_i, σ_i based on the conditions in Theorems 7.1 and 7.2; calculate the shortest path $d(i,j)$ linking all the possible subsystems i and j, then let $\overline{d} = \max\limits_{i,j \in \mathcal{M}} d(i,j)$.

On-line:

1: **for** $k=0,1,2,\ldots$ **do**

2: **if** $k=0$ **then**

3: For each controller i, set $t_i^0=0$, solve Problem 7.1 only with conditions (7.12)-(7.13) iteratively to obtain $\boldsymbol{u}_i^*(k+l \mid k), l=0,\ldots N-1$; apply $\boldsymbol{u}_i(k)=\boldsymbol{u}_i^*(k \mid k)$ to the system; broadcast $\boldsymbol{x}_i^*(k+l \mid k), l=1,\ldots,N$ to all its neighbors $j, j \in \mathcal{N}_i$. Set the switching signal $\theta_i=0$;

4: **else**

5: **while** $\boldsymbol{x}_i(k) \in \phi_i\left(\frac{\varepsilon}{2M}\right)$ **do**

6: Initialize $\theta_i=1$;

7: **for** $v=1:\overline{d}$ **do**

8: Receive θ_j from all neighbors $j \in \mathcal{N}_i$, update $\theta_i = \min\limits_{j \in \mathcal{N}_i} \{\theta_j, \theta_i\}$.

9: **end for**

10: **end while**

11: **if** $\theta_i=0$ **then**

12: For each controller i, at any time k, update $\hat{\boldsymbol{x}}_i(k+N \mid k)$ according to (7.14) or (7.15); calculate and check the triggering conditions (7.37) and (7.38);

13: **if** One of the triggering conditions in (7.39) is violated **then**

14: Solve the Problem 7.1 to obtain $\boldsymbol{u}_i^*(k+l \mid k), l=0,\ldots,N-1$; apply $\boldsymbol{u}_i(k)=\boldsymbol{u}_i^*(k \mid k)$ to the system; send $\boldsymbol{x}_i^*(k+l \mid k), l=1,\ldots,N$ to its neighbors $j, j \in \mathcal{N}_i$; set $r_i^{r+1}=k$;

15: **else**

16: Construct $\overline{\boldsymbol{u}}_i(k+l \mid k), l=0,\ldots,N-1$ based on (7.16) or (7.18); apply $\boldsymbol{u}_i(k)=\overline{\boldsymbol{u}}_i(k \mid k)$ to the system;

17: **end if**

18: **else**

All the controllers switch to the feedback control mode, apply $\boldsymbol{u}_i(k)=\boldsymbol{K}_{ii}\boldsymbol{x}_i(k)$.

end if

end if

end for

7.5 Feasibility and Stability Analysis

In this section, the recursive feasibility of the proposed Algorithm 1 and stability of the global system under it will be given.

Theorem 7.1 *For the NCS in (7.1), suppose that Assumption 7.1 holds, and there is a feasible solution initially. Problem 7.1 is feasible at every triggering instant if the following conditions hold:*

$$\frac{Nm_1}{\gamma\underline{\lambda}(P)}\sum_{h=0}^{N-2}\pi(N-h-2)\leqslant 1 \tag{7.34}$$

$$\delta_i\leqslant\frac{\kappa\varepsilon(1-\gamma)}{\max\limits_{l\in\{1,\ldots,N-1\}}2M\lambda^{-\frac{1}{2}}[\boldsymbol{P}_i^{-\frac{1}{2}}(\boldsymbol{A}_{ii}^l)^{\mathrm{T}}\boldsymbol{P}_i\boldsymbol{A}_{ii}^l\boldsymbol{P}_i^{-\frac{1}{2}}]} \tag{7.35}$$

$$1-\min_{i\in\mathcal{M}^-}\lambda(0.5\boldsymbol{P}_i^{-\frac{1}{2}}(\boldsymbol{Q}_i+\boldsymbol{K}_{ii}^{\mathrm{T}}\boldsymbol{R}_i\boldsymbol{K}_{ii})\boldsymbol{P}_i^{-\frac{1}{2}})\leqslant(1-\kappa)^2 \tag{7.36}$$

where $0<\gamma<1, 0<\kappa<1$.

Proof Suppose that Problem 7.1 is solved at t_i^r and the optimal control input $\boldsymbol{u}_i^*(t_i^r+l\,|t_i^r)$ and related optimal state $\boldsymbol{x}_i^*(t_i^r+l\,|t_i^r)$ are obtained. Next, we will prove that the constructed control candidate $\bar{\boldsymbol{u}}_i(k+l\,|k), l=0,\ldots,N-1$ at $k\in(t_i^r, t_i^{r+1}]$ in (7.14) and (7.16) satisfies all the constraints in Problem 7.1, where two cases are discussed: (a) For $k=t_i^r+1, \bar{\boldsymbol{u}}_i(k+l\,|k)$ in (7.14); (b) For $t_i^r+1<k\leqslant t_i^{r+1}, \bar{\boldsymbol{u}}_i(k+l\,|k)$ in (7.16). For brevity, only the case (b) is given as follows if unnecessary. Through a similar procedure, the same result can be obtained for case (a), where the main difference is that $\boldsymbol{u}_i^*(\cdot\,|t_i^r), \boldsymbol{x}_i^*(\cdot\,|t_i^r)$ are used instead of $\bar{\boldsymbol{u}}_i(\cdot\,|k-1), \bar{\boldsymbol{x}}_i(\cdot\,|k-1)$.

(i) $\sum_{h=0}^{l-1}\pi(l-h-1)\|\bar{\boldsymbol{x}}_i(k+h\,|k)-\boldsymbol{x}_i^*(k+h\,|t_i^r)\|\leqslant\frac{\gamma\kappa\varepsilon}{2Mm_1}$ holds for any $l=1,\ldots,N-1$.

(i-a) For $k = t_i^r + 1$, there is according to (7.21) that

$$\|\overline{\boldsymbol{x}}_i(k+l|k) - \boldsymbol{x}_i^*(k+l|t_i^r)\|_{\boldsymbol{P}_i}$$

$$\leqslant \|\boldsymbol{P}_i^{\frac{1}{2}}\boldsymbol{e}_i(k)\|_{\boldsymbol{P}_i^{-\frac{1}{2}}(\boldsymbol{A}_{ii}^l)^{\mathrm{T}}\boldsymbol{P}_i\boldsymbol{A}_{ii}^l\boldsymbol{P}_i^{-\frac{1}{2}}} + \sum_{j\in\mathcal{N}_i^u}\sum_{h=0}^{l-1}\|\boldsymbol{A}_{ii}^{l-h-1}\boldsymbol{A}_{ij}\eta_j[k+h|\tau_j(k)]\|_{\boldsymbol{P}_i}$$

$$\leqslant \overline{\lambda}^{\frac{1}{2}}[\boldsymbol{P}_i^{-\frac{1}{2}}(\boldsymbol{A}_{ii}^l)^{\mathrm{T}}\boldsymbol{P}_i\boldsymbol{A}_{ii}^l\boldsymbol{P}_i^{-\frac{1}{2}}]\|\boldsymbol{e}_i(k)\|_{\boldsymbol{P}_i}$$

$$+\sum_{j\in\mathcal{N}_i^u}\sum_{h=0}^{l-1}\pi(l-h-1)\|\eta_j[k+h|\tau_j(k)]\| \tag{7.37}$$

Let $r = arg \max_{j\in\mathcal{M}}\sum_{h=0}^{l-1}\pi(l-h-1)\|\eta_j[k+h|\tau_j(k)]\|$. Because $\eta_j[k+h|\tau_j(k)]$ is generated based on the optimal state sequence obtained at $\tau_j(k)$ where $\tau_j(k) < k$, it satisfies the constraints in the optimization problem, hence, it holds that

$$\sum_{h=0}^{l-1}\pi(l-h-1)\|\eta_r[k+h|\tau_j(k)]\leqslant\frac{\gamma\kappa\varepsilon}{2Mm_1} \tag{7.38}$$

Substituting (7.35) and (7.38) to (7.37), we get

$$\|\overline{\boldsymbol{x}}_i(k+l|k) - \boldsymbol{x}_i^*(k+l|t_i^r)\|_{\boldsymbol{P}_i}$$

$$\leqslant\overline{\lambda}^{\frac{1}{2}}[\boldsymbol{P}_i^{-\frac{1}{2}}(\boldsymbol{A}_{ii}^l)^{\mathrm{T}}\boldsymbol{P}_i\boldsymbol{A}_{ii}^l\boldsymbol{P}_i^{-\frac{1}{2}}]\delta_i + m_1\sum_{h=0}^{l-1}\pi(l-h-1)\|\eta_r[k+h|\tau_j(k)]\|$$

$$\leqslant\overline{\lambda}^{\frac{1}{2}}[\boldsymbol{P}_i^{-\frac{1}{2}}(\boldsymbol{A}_{ii}^l)^{\mathrm{T}}\boldsymbol{P}_i\boldsymbol{A}_{ii}^l\boldsymbol{P}_i^{-\frac{1}{2}})\delta_i + m_1\times\frac{\gamma\kappa\varepsilon}{2Mm_1}\leqslant\frac{(1-\gamma)\kappa\varepsilon}{2M}+\frac{\gamma\kappa\varepsilon}{2M}<\frac{\kappa\varepsilon}{2M} \tag{7.39}$$

According to the definition of $\pi(l-h-1)$ in (7.5), there is $\pi(l-h-1) > 0$. Hence by using (7.34), we obtain that for $l = 1, \ldots, N-1$,

$$\sum_{h=0}^{l-1}\pi(l-h-1)\|\overline{\boldsymbol{x}}_i(k+h|k) - \boldsymbol{x}_i^*(k+h|t_i^r)\|$$

$$\leqslant\sum_{h=0}^{l-1}\pi(l-h-1)\frac{\|\overline{\boldsymbol{x}}_i(k+h|k) - \boldsymbol{x}_i^*(k+h|t_i^r)\|_{\boldsymbol{P}_i}}{\underline{\lambda}(\boldsymbol{P}_i)} \tag{7.40}$$

$$\leqslant\sum_{h=0}^{l-1}\pi(l-h-1)\frac{m_1}{\gamma\underline{\lambda}(\boldsymbol{P}_i)}\times\frac{\gamma\kappa\varepsilon}{2Mm_1}$$

$$\leqslant\sum_{h=0}^{N-2}\pi(N-h-2)\frac{m_1}{\gamma\underline{\lambda}(\boldsymbol{P}_i)}\times\frac{\gamma\kappa\varepsilon}{2Mm_1}\leqslant\frac{\gamma\kappa\varepsilon}{2Mm_1}$$

Therefore, for $k=t_i^r+1$, constraint (7.7) is satisfied.

(i-b) For $t_i^r+1<k\leqslant t_i^{r+1}$, with a similar procedure, we get from (7.16) that for any $l=0,\ldots,N-2$, $\bar{\boldsymbol{u}}_i(k+l\,|\,k)=\bar{\boldsymbol{u}}_i(k+l\,|\,k-1)$. Then, based on (7.27), there is

$$
\begin{aligned}
&\|\bar{\boldsymbol{x}}_i(k+l\,|\,k)-\bar{\boldsymbol{x}}_i(k+l\,|\,k-1)\|_{\boldsymbol{P}_i}\\
&\leqslant\bar{\lambda}^{\frac{1}{2}}[\boldsymbol{P}_i^{-\frac{1}{2}}(\boldsymbol{A}_{ii}^l)^{\mathrm{T}}\boldsymbol{P}_i\boldsymbol{A}_{ii}^l\boldsymbol{P}_i^{-\frac{1}{2}}]\|\boldsymbol{e}_i(k)\|_{\boldsymbol{P}_i}\\
&+\sum_{j\in\mathcal{N}_i^u}\sum_{h=0}^{l-1}\pi(l-h-1)\|\boldsymbol{x}_j^*[k+h\,|\,\tau_j(k)]-\boldsymbol{x}_j^*[k+h\,|\,\tau_j(k-1)]\|\\
&\leqslant\bar{\lambda}^{\frac{1}{2}}[\boldsymbol{P}_i^{-\frac{1}{2}}(\boldsymbol{A}_{ii}^l)^{\mathrm{T}}\boldsymbol{P}_i\boldsymbol{A}_{ii}^l\boldsymbol{P}_i^{-\frac{1}{2}}]\delta_i+m_1\times\frac{\gamma\kappa\varepsilon}{2Mm_1}\leqslant\frac{(1-\gamma)\kappa\varepsilon}{2M}+\frac{\gamma\kappa\varepsilon}{2M}<\frac{\kappa\varepsilon}{2M}
\end{aligned}
\tag{7.41}
$$

Thus, we have

$$
\begin{aligned}
&\|\bar{\boldsymbol{x}}_i(k+l\,|\,k)-\boldsymbol{x}_i^*(k+l\,|\,t_i^r)\|_{\boldsymbol{P}_i}\\
&\leqslant\|\bar{\boldsymbol{x}}_i(k+l\,|\,k)-\bar{\boldsymbol{x}}_i(k+l\,|\,k-1)\|_{\boldsymbol{P}_i}+\cdots+\|\bar{\boldsymbol{x}}_i(k+l\,|\,t_i^r+1)-\bar{\boldsymbol{x}}_i(k+l\,|\,t_i^r)\|_{\boldsymbol{P}_i}\\
&\leqslant\frac{N\kappa\varepsilon}{2M}
\end{aligned}
\tag{7.42}
$$

With the same as (7.40), we obtain that

$$
\begin{aligned}
&\sum_{h=0}^{l-1}\pi(l-h-1)\|\bar{\boldsymbol{x}}_i(k+h\,|\,k)-\boldsymbol{x}_i^*(k+h\,|\,t_i^r)\|\\
&\leqslant\sum_{h=0}^{l-1}\pi(l-h-1)\frac{\|\bar{\boldsymbol{x}}_i(k+h\,|\,k)-\boldsymbol{x}_i^*(k+h\,|\,t_i^r)\|_{\boldsymbol{P}_i}}{\underline{\lambda}(\boldsymbol{P}_i)}\\
&\leqslant\sum_{h=0}^{l-1}\pi(l-h-1)\frac{Nm_1}{\gamma\underline{\lambda}(P_i)}\times\frac{\gamma\kappa\varepsilon}{2Mm_1}\\
&\leqslant\sum_{h=0}^{N-2}\pi(N-h-2)\frac{Nm_1}{\gamma\underline{\lambda}(\boldsymbol{P}_i)}\times\frac{\gamma\kappa\varepsilon}{2Mm_1}\leqslant\frac{\gamma\kappa\varepsilon}{2Mm_1}
\end{aligned}
\tag{7.43}
$$

Summarizing (i-a) and (i-b), it completes the proof.

(ii) $\|\bar{\boldsymbol{x}}_i(k+N\,|\,k)-\hat{\boldsymbol{x}}_i(k+N\,|\,k)\|_{\boldsymbol{P}_i}\leqslant\frac{\kappa\varepsilon}{2M}$.

For $k>t_i^r+1$, based on (7.28) and Lemma 7.1, it follows from (7.41) that

$$\begin{aligned}&\|\overline{\boldsymbol{x}}_i(k+N|k)-\hat{\boldsymbol{x}}_i(k+N|k)\|_{\boldsymbol{P}_i}\\&=\|\boldsymbol{A}_{d_{ii}}[\overline{\boldsymbol{x}}_i(k+N-1|k)-\overline{\boldsymbol{x}}_i(k+N-1|k-1)]\|_{\boldsymbol{P}_i}\\&\leqslant\|\overline{\boldsymbol{x}}_i(k+N-1|k)-\overline{\boldsymbol{x}}_i(k+N-1|k-1)\|_{\boldsymbol{P}_i}\leqslant\frac{\kappa\varepsilon}{2M}\end{aligned}\tag{7.44}$$

This completes the proof.

(iii) $\overline{\boldsymbol{x}}_i(k+N|k)\in\phi_i(\frac{\varepsilon}{2M})$.

It follows from (ii) that

$$\begin{aligned}&\|\overline{\boldsymbol{x}}_i(k+N|k)\|_{\boldsymbol{P}_i}\\&\leqslant\|\overline{\boldsymbol{x}}_i(k+N|k)-\hat{\boldsymbol{x}}_i(k+N|k)\|_{\boldsymbol{P}_i}+\|\hat{\boldsymbol{x}}_i(k+N|k)\|_{\boldsymbol{P}_i}\\&\leqslant\frac{\kappa\varepsilon}{2M}+\|\hat{\boldsymbol{x}}_i(k+N|k)\|_{\boldsymbol{P}_i}\end{aligned}\tag{7.45}$$

For $k>t_i^r+1$, using Lemma 7.1 and $\overline{\boldsymbol{x}}_i(k+N-1|k-1)\in\phi_i(\frac{\varepsilon}{2M})$, we obtain from (7.13) that

$$\begin{aligned}&\|\hat{\boldsymbol{x}}_i(k+N|k)\|^2_{\boldsymbol{P}_i}-\|\overline{\boldsymbol{x}}_i(k+N-1|k-1)\|^2_{\boldsymbol{P}_i}\\&\leqslant-\|\overline{\boldsymbol{x}}_i(k+N-1|k-1)\|^2_{\boldsymbol{P}_i^{-\frac{1}{2}}\hat{\boldsymbol{Q}}_i\boldsymbol{P}_i^{-\frac{1}{2}}}\\&\leqslant-\underline{\lambda}(\boldsymbol{P}_i^{-\frac{1}{2}}\hat{\boldsymbol{Q}}_i\boldsymbol{P}_i^{-\frac{1}{2}})\|\overline{\boldsymbol{x}}_i(k+N-1|k-1)\|^2_{\boldsymbol{P}_i}\end{aligned}\tag{7.46}$$

where $\hat{\boldsymbol{Q}}_i=(\boldsymbol{Q}_i+\boldsymbol{K}_i^{\mathrm{T}}\boldsymbol{R}_i\boldsymbol{K}_i)/2$. It is obvious that if condition (7.36) is met, there is

$$\begin{aligned}\|\hat{\boldsymbol{x}}_i(k+N|k)\|_{\boldsymbol{P}_i}&\leqslant\sqrt{1-\underline{\lambda}(\boldsymbol{P}_i^{-\frac{1}{2}}\hat{\boldsymbol{Q}}_i\boldsymbol{P}_i^{-\frac{1}{2}})}\|\overline{\boldsymbol{x}}_i(k+N-1|k-1)\|_{\boldsymbol{P}_i}\\&\leqslant\sqrt{1-\underline{\lambda}(\boldsymbol{P}_i^{-\frac{1}{2}}\hat{\boldsymbol{Q}}_i\boldsymbol{P}_i^{-\frac{1}{2}})}\times\frac{\varepsilon}{2M}\leqslant\frac{(1-\kappa)\varepsilon}{2M}\end{aligned}\tag{7.47}$$

Substituting (7.47) to (7.45), we have

$$\|\overline{\boldsymbol{x}}_i(k+N|k)\|_{\boldsymbol{P}_i}\leqslant\frac{\kappa\varepsilon}{2M}+\frac{(1-\kappa)\varepsilon}{2M}=\frac{\varepsilon}{2M}\tag{7.48}$$

Therefore, $\overline{\boldsymbol{x}}_i(k+N|k)$ can be driven into the terminal set at k.

(iv) $\overline{\boldsymbol{u}}_i(k+l|k)\in\mathcal{U}_i$, $l=0,\ldots,N-1$.

For $k>t_i^r+1$, according to (7.16), we get $\bar{\boldsymbol{u}}_i(k+l|k)=\bar{\boldsymbol{u}}_i(k+l|k-1)\in \mathcal{U}_i, l=0,\ldots,N-2$.

It further follows from (7.41) that

$$\begin{aligned}
&\|\bar{\boldsymbol{x}}_i(k+N-1|k)\|_{\boldsymbol{P}_i}\\
&\leqslant\|\bar{\boldsymbol{x}}_i(k+N-1|k)-\bar{\boldsymbol{x}}_i(k+N-1|k-1)\|_{\boldsymbol{P}_i}+\|\bar{\boldsymbol{x}}_i(k+N-1|k-1)\|_{\boldsymbol{P}_i}\\
&\leqslant\frac{\kappa\varepsilon}{2M}+\frac{\varepsilon}{2M}<\frac{\varepsilon}{M}
\end{aligned}\tag{7.49}$$

Since $\bar{\boldsymbol{x}}_i(k+N-1|k)\in\phi_i(\varepsilon/M)$, it is obtained from Assumption 7.1 that $\bar{\boldsymbol{u}}_i(k+N-1|k)=\boldsymbol{K}_{ii}\bar{\boldsymbol{x}}_i(k+N-1|k)\in\mathcal{U}_i$.

Therefore, $\bar{\boldsymbol{u}}_i(k+l|k)\in\mathcal{U}_i$ for any $l=0,\ldots,N-1$.

(v) It is easily found that $\|\bar{\boldsymbol{x}}_i(k+l|k)\|_{\boldsymbol{P}_i}\leqslant\|\bar{\boldsymbol{x}}_i(k+l|k)\|_{\boldsymbol{P}_i}+\frac{\varepsilon}{\rho NM}$ holds all the time.

Summarizing (i)-(v), the constructed control candidate $\bar{\boldsymbol{u}}_i(k+l|k), l=0,\ldots,N-1$ is feasible to Problem 7.1. This completes the proof. □

Remark 7.4 From (7.34), it is found that the prediction horizon is constrained by γ. Specifically speaking, a larger value of γ causes a wider range available for N. Besides, it is derived from constraint (7.35) that as κ goes up and γ goes down, the upper bound on δ_i becomes larger, which indicates that the average period of communication may become longer and thus communication cost is more reduced. As for the margin $\frac{\gamma\kappa\varepsilon}{2Mm_1}$ in constraint (7.7), it will be increased as κ and γ become larger. According to (7.36), an upper bound on κ is obtained.

Theorem 7.2 *For the system in (7.2), suppose Assumption 7.1 holds and all the conditions in Theorem 7.1 are satisfied. The overall system under the proposed Algorithm 1 is stabilizing if the selected event-triggering threshold δ_i and prediction horizon N are also satisfied the following condition:*

$$\sum_{i=1}^{M}\delta_i+\frac{(N-1)\kappa\varepsilon}{2}+\frac{\varepsilon}{\rho}<\frac{\varepsilon}{2}\tag{7.50}$$

Proof In the proposed Algorithm 1, the dual-mode strategy is applied and thus two Lyapunov candidate functions are defined: (a) for $\boldsymbol{x}(k) \notin \phi(\varepsilon)$, $\boldsymbol{V}_{out}[\boldsymbol{x}(k)] = \sum_{i=1}^{M} \boldsymbol{V}_i[\boldsymbol{x}_i(\,|\,k)] = \sum_{i=1}^{M}\sum_{l=0}^{N} \|\boldsymbol{x}_i(k+l\,|\,k)\|_{\boldsymbol{P}_i}$; (b) for $\boldsymbol{x}(k) \in \phi(\varepsilon)$, $\boldsymbol{V}_{in}[\boldsymbol{x}(k)] = \|\boldsymbol{x}(k)\|_{\boldsymbol{P}}$. According to the Theorem 2.7 in [20], the global system is stable if the following two conditions are satisfied: (1) $\boldsymbol{V}_{out}[\boldsymbol{x}(k)]$ and $\boldsymbol{V}_{in}[\boldsymbol{x}(k)]$ are monotonically decreasing; (2) $\boldsymbol{V}_{in}[\boldsymbol{x}(l)] - \boldsymbol{V}_{out}[\boldsymbol{x}(l-1)] < 0$ if the switching signal happens at time instant l. It follows from Lemma 7.1 that $\boldsymbol{V}_{in}[\boldsymbol{x}(k)]$ decreases as time goes on, and it further has that

$$
\begin{aligned}
\boldsymbol{V}_{in}[\boldsymbol{x}(l)] - \boldsymbol{V}_{out}[\boldsymbol{x}(l-1)] &= \|\boldsymbol{x}(l)\|_{\boldsymbol{P}} - \sum_{i=1}^{M} \|\boldsymbol{x}_i(k+l\,|\,k)\|_{\boldsymbol{P}_i} \\
&\quad - \sum_{i=1}^{M}\sum_{l=1}^{N} \|\boldsymbol{x}_i(k+l\,|\,k)\|_{\boldsymbol{P}_i} \\
&\leqslant \|\boldsymbol{x}(l)\|_{\boldsymbol{P}} - \|\boldsymbol{x}(l-1)\|_{\boldsymbol{P}} - \sum_{i=1}^{M}\sum_{l=1}^{N} \|\boldsymbol{x}_i(k+l\,|\,k)\|_{\boldsymbol{P}_i} \\
&< -\sum_{i=1}^{M}\sum_{l=1}^{N} \|\boldsymbol{x}_i(k+l\,|\,k)\|_{\boldsymbol{P}_i} < 0
\end{aligned}
\tag{7.51}
$$

In the following, $\boldsymbol{V}_{out}[\boldsymbol{x}(k)] - \boldsymbol{V}_{out}[\boldsymbol{x}(k-1)] < 0$ will be proved to show the stability of the closed-loop system.

Firstly, since the constructed feasible control inputs $\overline{\boldsymbol{u}}_i(k+l\,|\,k)$ are applied if there is no event occurring at k, it follows from (7.24) and (7.29) that

$$
\boldsymbol{V}_i[\boldsymbol{x}_i(k)] - \boldsymbol{V}_i[\boldsymbol{x}_i(k-1)] \leqslant (\sigma_i - 1)\|\boldsymbol{x}_i(k-1)\|_{\boldsymbol{P}_i} < 0
$$

Otherwise if either of the constructed event-triggering conditions is violated at k, Problem 7.1 is solved, then $\boldsymbol{u}_i^*(k+l\,|\,k)$ and $\boldsymbol{x}_i^*(k+l\,|\,k)$ are obtained, which satisfy constraint (7.9):

$$
\|\boldsymbol{x}_i^*(k+l\,|\,k)\|_{\boldsymbol{P}_i} \leqslant \|\overline{\boldsymbol{x}}_i(k+l\,|\,k)\|_{\boldsymbol{P}_i} + \frac{\varepsilon}{\rho NM}, l=1,\dots,N
$$

Then the difference $\boldsymbol{V}_i[\boldsymbol{x}_i^*(k+l\,|\,k)] - \boldsymbol{V}_i[\overline{\boldsymbol{x}}_i(k+l\,|\,k-1)]$ is

$$
\begin{aligned}
&\boldsymbol{V}_i[\boldsymbol{x}_i^*(k\,|\,k)] - \boldsymbol{V}_i[\overline{\boldsymbol{x}}_i(k-1\,|\,k-1)] \\
&\leqslant \frac{\varepsilon}{\rho M} + \sum_{l=0}^{N} \|\overline{\boldsymbol{x}}_i(k+l\,|\,k)\|_{\boldsymbol{P}_i} - \sum_{l=0}^{N} \|\overline{\boldsymbol{x}}_i(k+l-1\,|\,k-1)\|_{\boldsymbol{P}_i}
\end{aligned}
$$

$$\leqslant \frac{\varepsilon}{\rho M} + \sum_{l=1}^{N-1} \| \overline{\boldsymbol{x}}_i(k+l|k) - \overline{\boldsymbol{x}}_i(k+l|k-1) \|_{\boldsymbol{P}_i} + \| \overline{\boldsymbol{x}}_i(k+N|k) \|_{\boldsymbol{P}_i}$$
$$+ \| \boldsymbol{x}_i(k) - \overline{\boldsymbol{x}}_i(k|k-1) \|_{\boldsymbol{P}_i} - \| \boldsymbol{x}_i(k-1) \|_{\boldsymbol{P}_i} \tag{7.52}$$

It follows from (i)and (iii) in the feasibility analysis that

$$\boldsymbol{V}_i[\boldsymbol{x}_i^*(k|k)] - \boldsymbol{V}_i[\overline{\boldsymbol{x}}_i(k-1|k-1)] \leqslant \frac{\varepsilon}{\rho M} + \frac{(N-1)\kappa\varepsilon}{2M} + \frac{\varepsilon}{2M} + \delta_i - \| \boldsymbol{x}_i(k-1) \|_{\boldsymbol{P}_i} \tag{7.53}$$

With a similar procedure, the difference between $\boldsymbol{V}_i[\boldsymbol{x}_i^*(k|k)]$ and $\boldsymbol{V}_i[\boldsymbol{x}_i^*(k-1|k-1)]$ is obtained if the controller is activated at $k-1$, and it is identical with (7.53). Since the difference of V_i at k and $k-1$ is in a same form whether an event occurs or not, $\boldsymbol{x}_i^*(k+l|k)$ or $\overline{\boldsymbol{x}}_i(k+l|k)$ is replaced by $\boldsymbol{x}_i(k+l|k)$ in the following for brevity.

Finally, $\boldsymbol{V}_{out}[\boldsymbol{x}(k)] - \boldsymbol{V}_{out}[\boldsymbol{x}(k-1)]$ is calculated by

$$\boldsymbol{V}_{out}[\boldsymbol{x}(k)] - \boldsymbol{V}_{out}[\boldsymbol{x}(k-1)] = \sum_{i=1}^{M} \{ \boldsymbol{V}_i[\boldsymbol{x}_i(k|k)] - \boldsymbol{V}_i[\boldsymbol{x}_i(k-1|k-1)] \}$$
$$\leqslant \sum_{i=1}^{M} \left[\frac{\varepsilon}{\rho M} + \frac{(N-1)\kappa\varepsilon}{2M} + \frac{\varepsilon}{2M} + \delta_i - \| \boldsymbol{x}_i(k-1) \|_{\boldsymbol{P}_i} \right]$$
$$\leqslant \frac{\varepsilon}{\rho} + \frac{\varepsilon}{2} + \frac{(N-1)\kappa\varepsilon}{2} + \sum_{i=1}^{M} \delta_i - \| \boldsymbol{x}(k-1) \|_{\boldsymbol{P}}$$

Since $\boldsymbol{x}(k-1) \notin \phi(\varepsilon)$, thus

$$\boldsymbol{V}_{out}[\boldsymbol{x}(k)] - \boldsymbol{V}_{out}[(k-1)] \leqslant \frac{\varepsilon}{\rho} + \frac{\varepsilon}{2} + \frac{(N-1)\kappa\varepsilon}{2} + \sum_{i=1}^{M} \delta_i - \varepsilon$$
$$\leqslant \sum_{i=1}^{M} \delta_i + \frac{(N-1)\kappa\varepsilon}{2} + \frac{\varepsilon}{\rho} - \frac{\varepsilon}{2}$$

If (7.50)holds, there is $\boldsymbol{V}_{out}[\boldsymbol{x}(k)] - \boldsymbol{V}_{out}[(k-1)] < 0$. Therefore, the system in (7.2)can be driven to the terminal set $\phi(\varepsilon)$ in a finite time under the proposed event-triggered DMPC strategy. □

Remark 7.5 As ρ goes up, the decay rate of the defined Lyapunov function increases, but the margin in constraint (7.9) for stability guarantees is reduced.

7.6 Example

In this section, simulations on an numerical model and a four-tank plant under the proposed Algorithm 1 (denoted as ET-DMPC) are made to verify its effectiveness.

Example 1 Consider a NCS composed of four subsystems with state coupling, where the system matrices of the overall system is given by

$$\boldsymbol{A}=\begin{bmatrix}1.24 & -0.2 & 0 & 0\\ 0 & 0.91 & 0 & 0\\ 0.13 & -0.27 & 1.06 & 0\\ 0 & 0 & 0.23 & 1.14\end{bmatrix},\boldsymbol{B}=\begin{bmatrix}0.24\\ 0.23\\ 0.24\\ 0.26\end{bmatrix}$$

The control input is bounded by $-1\leqslant \boldsymbol{u}_i\leqslant 1, i=1,2,3,4$. Our control goal is to drive system state to the equilibrium from $x_1(0)=0.74, x_2(0)=0.78, x_3(0)=0.6, x_4(0)=0.5$.

The parameters in the cost function are selected according to the Assumption 7.1 and Lemma 7.1. Specifically, the prediction horizon is set $N=3$, the weighting matrices are chosen as $\boldsymbol{Q}_i=6, \boldsymbol{R}_i=1$ for any $i=1,2,3,4$, and the feedback gains are calculated by solving LMIs [18] and given by $K_{11}=-3.5199$, $K_{22}=-2.1568, K_{33}=-2.702, K_{44}=-2.7781$. The terminal matrices $\boldsymbol{P}_i, i=1,2,3,4$ are selected to satisfy Lemma 7.1 and the conditions in Theorem 7.1, which are set as $P_1=24.4351, P_2=15.3769, P_3=18.019, P_4=18.7368$ with $\varepsilon=0.6$. To ensure recursive feasibility of the proposed algorithm, the lower bound of γ is obtained based on the condition (7.34): $0.6545\leqslant\gamma<1$, the upper bound of κ is calculated according to the condition (7.36): $0<\kappa<0.1807$. With considering the feasibility and stability properties, the triggering threshold is evaluated based on the main design parameters: $0\leqslant\delta_1\leqslant 0.0022, 0\leqslant\delta_2\leqslant 0.0037, 0\leqslant\delta_3\leqslant 0.003, 0\leqslant\delta_4\leqslant 0.0026$. By adjusting the weighting factors σ_i, $0\leqslant\sigma_i<1$ in the event-triggering condition, it achieves different trade-offs between control performance and communication cost.

In the implementation, we choose $\gamma=0.7, \kappa=0.18, \rho=8, \sigma_1=0.9, \sigma_2=0.05, \sigma_3=0.5, \sigma_4=0.9, w_1=0.2, w_2=0.5, w_3=0.9, w_4=0$, and carry out

Algorithm 1. The corresponding results are depicted in Figs. 7. 2, 7. 3, 7. 4, 7. 5 and 7. 6. In Figs. 7. 2 and 7. 3, the state trajectories and control inputs of subsystems 1-4 are shown, respectively. It is easily found that the global system is stabilizing and all the control inputs meet their constraints. It obtains from Fig. 7. 4 that the local subsystem approaches to its terminal set from step 2 to step 8, and the global system is driven into the terminal region at step $k=5$. At this step, all the local controllers are switched from the event-triggered MPC strategy to the state feedback control mode. In Fig. 7. 5, the exact triggering instants of the four subsystems are shown, where "1" denotes that an event occurs, thus Problem 7. 1 is solved and the state information is sent, "0" means that no event happens and the constructed feasible control input is applied. In order to quantitatively evaluate the saved communication cost, denote the total number of the events as s and define the average triggering period of subsystem i as $\overline{T}_i$, then it has

$$\overline{T}_i=\sum_{h=0}^{s-2}\frac{t_i^{h+1}-t_i^h}{s-1}=\frac{t_i^{s-1}}{s-1} \tag{7.54}$$

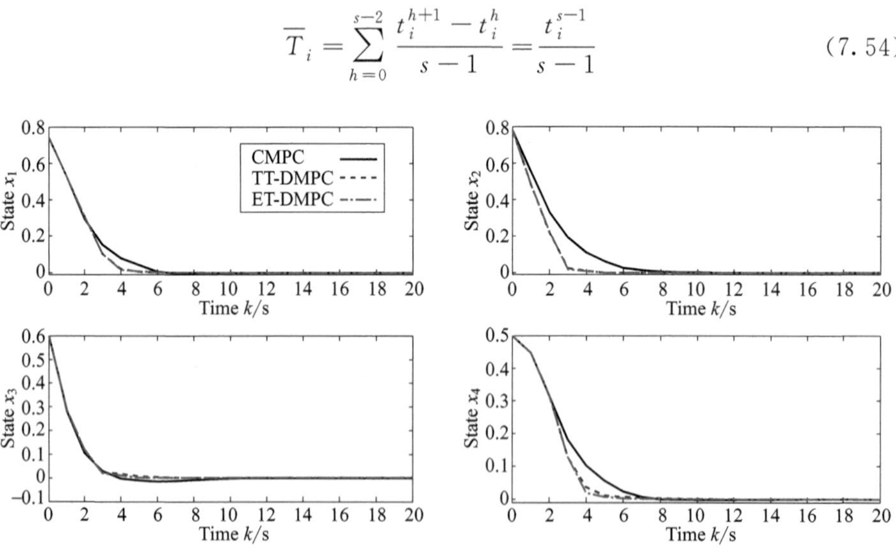

Fig. 7. 2 State trajectories of subsystems 1-4 using CMPC, TT-DMPC and ET-DMPC strategies

For the four controllers, their average triggering periods are calculated: $\overline{T}_1=2\text{s}$, $\overline{T}_2=\overline{T}_3=3\text{s}$, $\overline{T}_4=1\text{s}$, which indicates that the four controllers aperiodically and asynchronously solve Problem 7. 1 and coordinate with each other, thus reducing the cost of computation and communication.

To show the effectiveness of the proposed ET-DMPC, the centralized MPC

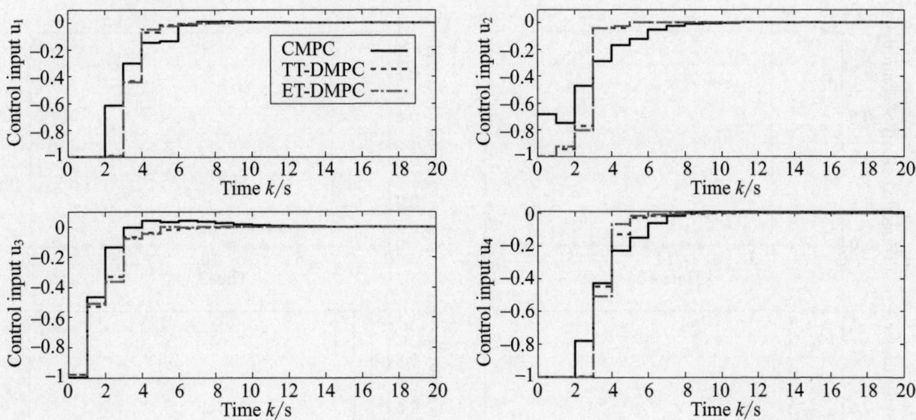

Fig. 7.3 Control inputs of subsystems 1-4 using CMPC, TT-DMPC and ET-DMPC strategies

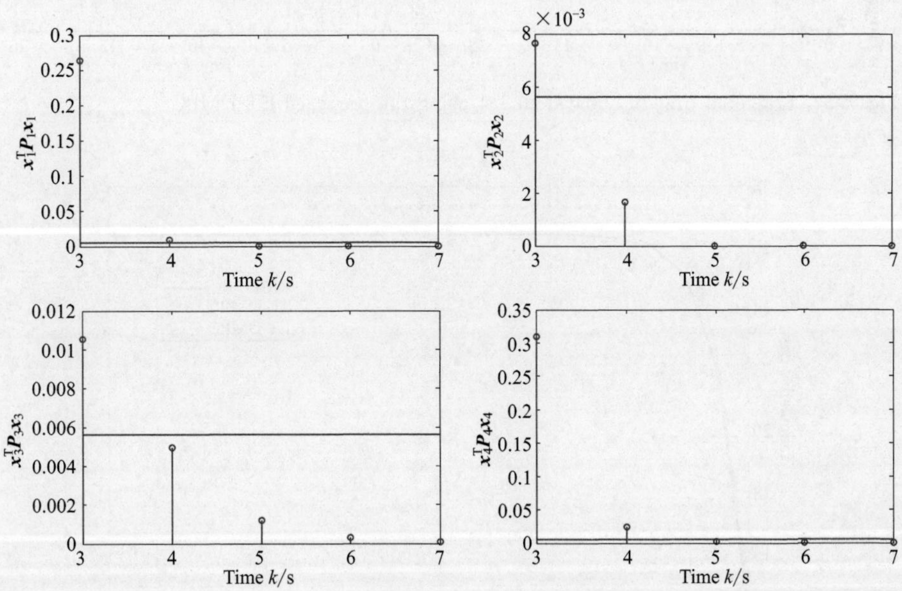

Fig. 7.4 $x_i^{\mathrm{T}} P_i x_i$ of subsystem 1-4 using the proposed ET-DMPC

(denoted by CMPC) and DMPC are also carried out with the same parameters but in a time-triggered manner. Figs. 7.2 and 7.3 presents a comparison on the state trajectories and control inputs of the four subsystems. In order to quantitatively evaluate the loss/gain of control performance, a cumulated cost is considered and defined by

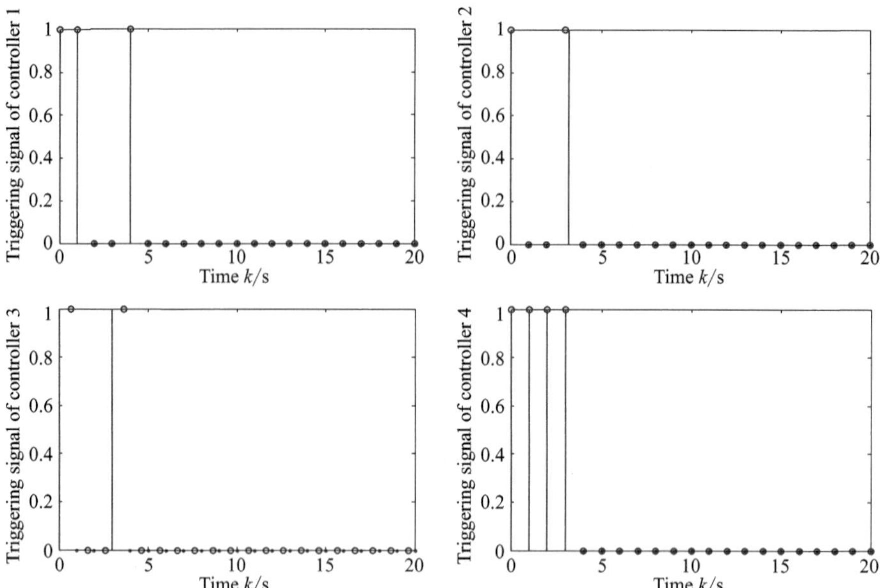

Fig. 7. 5 Triggered time of subsystem 1-4 using the proposed ET-DMPC

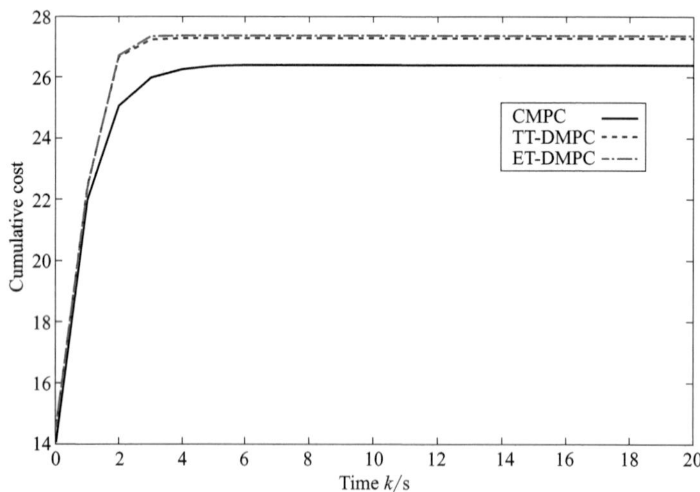

Fig. 7. 6 Cumulative costs using CMPC, TT-DMPC and ET-DMPC

$$Jcc(k)=\sum_{i\in\mathcal{M}}\sum_{l=0}^{k}\|\boldsymbol{x}_i(k)\|_{Q_i}^2+\|\boldsymbol{u}_i(k)\|_{\boldsymbol{R}_i}^2, k=0,\ldots,Nsim \tag{7.55}$$

where $N\ sim$ denotes the simulation steps. Fig. 7. 6 depicts the trajectories of Jcc under CMPC, TT-DMPC and ET-DMPC, which converge to 26. 3956, 27. 2711 and 27. 3584, respectively. Obviously, the control performance under

ET-DMPC is extremely close to the one achieved under TT-DMPC, but a little poorer than the one under CMPC.

In conclusion, a satisfied control performance can be achieved while reducing the frequency of communication and computation under the proposed ET-DMPC.

Example 2 The four-tank system shown in Fig. 7.7 is considered, which is a benchmark case used to verify the effectiveness of distributed control algorithm. For any tank $i=1,2,3,4$, h_i, A_i and a_i are its water level, cross-section and the cross-section of the outlet hole, respectively. For each pump i, $i=1,2$, v_i and k_i are its voltage and the conversion parameter from its voltage to the flux of water respectively. For the three-way valve $i=1,2$, its ration is denoted respectively by β_1 and β_2. Besides, g is the gravitational acceleration. Their exact values are taken from [19]: $A_1=A_4=28\text{cm}^2$, $A_2=A_3=32\text{cm}^2$, $a_1=a_4=0.071\text{cm}^2$, $a_2=a_3=0.057\text{cm}^2$, $k_1=3.35\text{cm}^3/\text{Vs}$, $k_2=3.33\text{cm}^3/\text{Vs}$, $\beta_1=0.7$, $\beta_2=0.6$. Here, our control objective is to manage the water levels h_i of the four tanks to their operating points by adjusting the two pumps' voltages v_i under the proposed ET-DMPC. Firstly, the nonlinear model of the four-tank system is linearized and discretized with sampling time $T=1$ s at their operating points $h_1^0=12.263$ cm, $h_2^0=1.409$cm, $h_3^0=12.783$cm, $h_4^0=1.634$ cm, $v_1^0=v_2^0=3$V. Denote $\delta h_i=h_i-h_i^0$, $i=1,2,3,4$ and $\delta v_i=v_i-v_i^0$, $i=1,2$, and is described by

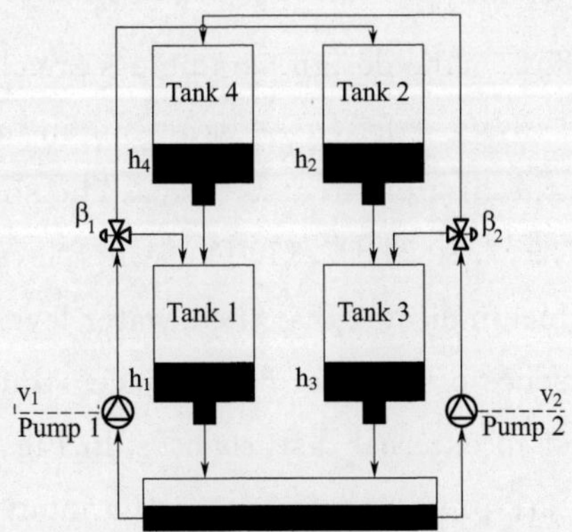

The simplified model of this quadruple-tank process is given by

$$\frac{dh_1}{dt}=-\frac{a_1}{A_1}\sqrt{2gh_1}+\frac{a_4}{A_4}\sqrt{2gh_4}+\frac{\beta_1 k_1}{A_1}v_1$$

$$\frac{dh_2}{dt}=-\frac{a_2}{A_2}\sqrt{2gh_2}+\frac{(1-\beta_1)k_1}{A_2}v_1$$

$$\frac{dh_3}{dt}=-\frac{a_3}{A_3}\sqrt{2gh_3}+\frac{a_2}{A_2}\sqrt{2gh_2}+\frac{\beta_2 k_2}{A_3}v_2$$

$$\frac{dh_4}{dt}=-\frac{a_4}{A_4}\sqrt{2gh_4}+\frac{(1-\beta_2)k_2}{A_4}v_2$$

Fig. 7.7 The four-tank process diagram

$$\begin{bmatrix}\delta h_1(k+1)\\ \delta h_2(k+1)\\ \delta h_3(k+1)\\ \delta h_4(k+1)\end{bmatrix}=\begin{bmatrix}0.98 & 0 & 0 & 0.04\\ 0 & 0.97 & 0 & 0\\ 0 & 0.03 & 0.99 & 0\\ 0 & 0 & 0 & 0.96\end{bmatrix}\begin{bmatrix}\delta h_1(k)\\ \delta h_2(k)\\ \delta h_3(k)\\ \delta h_4(k)\end{bmatrix}+\begin{bmatrix}0.08 & 0\\ 0.03 & 0\\ 0 & 0.06\\ 0 & 0.05\end{bmatrix}\begin{bmatrix}\delta v_1(k)\\ \delta v_2(k)\end{bmatrix} \tag{7.56}$$

Divide the four-tank system into two coupled subsystems, where subsystem 1 makes up of tanks 1 and 2 and subsystem 2 consists of tanks 3 and 4. Then further define the inputs and states of subsystems 1 and 2 as $\boldsymbol{x}_1=[\delta h_1,\delta h_2]^{\mathrm{T}}$, $\boldsymbol{x}_2=[\delta h_3,\delta h_4]^{\mathrm{T}}$, $\boldsymbol{u}_1=\delta v_1$, $\boldsymbol{u}_2=\delta v_2$ respectively, and their dynamic behaviors are derived from (7.56). For any subsystem $i=1,2$, their control inputs are limited by $-3\leqslant \boldsymbol{u}_i\leqslant 3$.

In the design of each controller $i, i=1,2$, its weighting matrices are chosen as $\boldsymbol{Q}_i=\boldsymbol{I}_2, \boldsymbol{R}_i=1$. The feedback gains $\boldsymbol{K}_i$ and terminal matrix $\boldsymbol{P}_i$ are obtained by solving LMI [18], and given by

$$\boldsymbol{K}_1=[-1.0439\ -0.2726], \boldsymbol{K}_2=[-1.2131\ -0.3562],$$

$$\boldsymbol{P}_1=\begin{bmatrix}19.0046 & -12.2731\\ -12.2731 & 44.1559\end{bmatrix}, \boldsymbol{P}_2=\begin{bmatrix}34.8631 & -15.2952\\ -15.2952 & 26.9819\end{bmatrix}$$

The terminal set level is set as $\varepsilon=0.6$ and the prediction horizon are chosen as $N=8$. Then according to the conditions in Theorems 7.1 and 7.2, the bounds on other design parameters are obtained: $0.6872<\gamma<1$, $0<\kappa<0.0051$, $0<\delta_1<7.343\times10^{-4}$, $0<\delta_2<7.256\times10^{-4}$.

In implementation, the initial conditions are set as $\boldsymbol{x}_1(0)=[0.274\quad 0.167]^{\mathrm{T}}$, $\boldsymbol{x}_2(0)=[0.203\quad 0.254]^{\mathrm{T}}$, the design parameters are chosen as $\kappa=0.0048$, $\gamma=0.8$, $\sigma_1=0.35$, $\sigma_2=0.45$, $w_1=0.3$, $w_2=0.2$, and an unpredicted impulse equal to 2V is forced to the first pump at $k=100$. The simulation results are depicted in Figs. 7.8, 7.9, 7.10 and 7.11. The water levels of the four tanks are shown in Fig. 7.8, which indicates that their water levels are driven to the desired point, even in presence of external disturbances, thus the proposed ET-DMPC is somewhat robust to external disturbances. In Fig. 7.9, the applied voltages to the two pumps are presented and meet the input constraints. The triggering time steps of the two controllers are illustrated in Fig. 7.10, where "1" means that the controller is activated at this step, "0" indicates that no event happens. It is found that the two controllers are switched from the event-triggered DMPC strategy to the state feedback control mode at $k=16$. Their average triggering periods are $\overline{T}_1=3.75$s and $\overline{T}_2=3.3$s respectively.

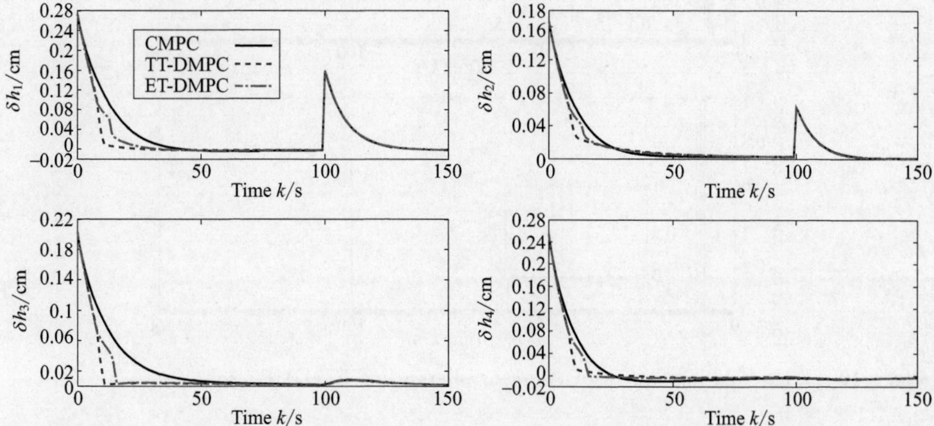

Fig. 7.8 Water levels δh_i of four tanks using CMPC, TT-DMPC and ET-DMPC

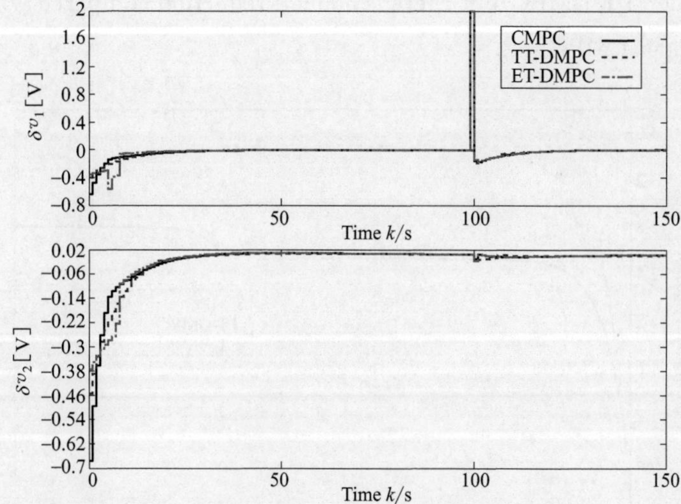

Fig. 7.9 Voltages δv_1, δv_2 under CMPC, TT-DMPC and ET-DMPC

The CMPC and TT-DMPC algorithms with the same parameters are also adopted to the four-tank system. The corresponding simulation results on water levels and the voltages are shown in Figs. 7.8 and 7.9 respectively. The trajectories of cumulative costs Jcc under CMPC, TT-DMPC and ET-DMPC

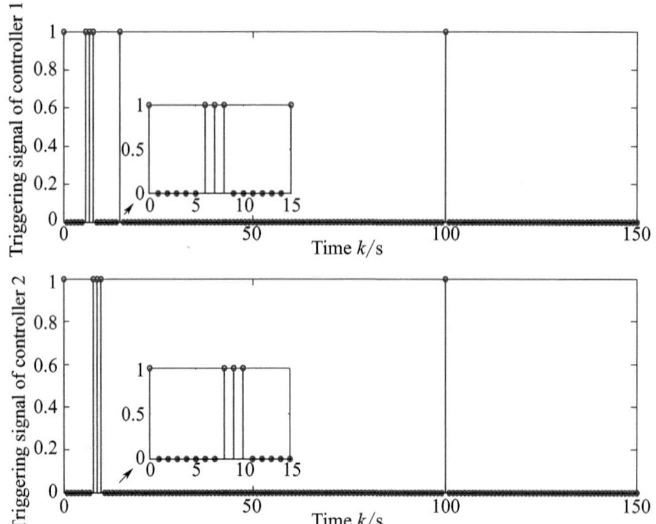

Fig. 7. 10 Triggering instants of of two controllers using ET-DMPC scheme

are presented in Fig. 7. 11, which converge to 2. 8184, 3. 7781 and 3. 2846 respectively. It is found that control performance under ET-DMPC is better than the one under TT-DMPC, but there is a small reduction compared with the one achieved under CMPC.

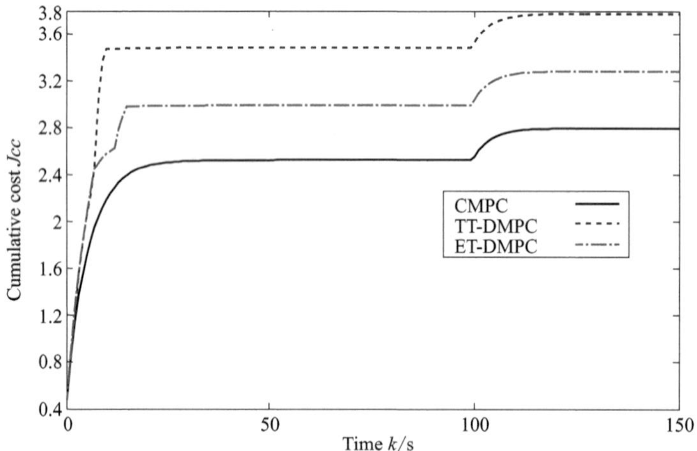

Fig. 7. 11 Cumulative costs using CMPC, TT-DMPC and ET-DMPC

Therefore, the proposed ET-DMPC algorithm is capable to keep a good control performance while reducing the communication and computation load.

7.7 Conclusion

In this chapter, the problem of an event-triggered DMPC for the physically interconnected NCSs has been investigated. Considering the mutual influences, the DMPC optimization control problems for the event-triggered subsystems have been established, where an approximated control performance of its downstream neighbors is considered in the cost function of each local subsystem. Then two triggering event-triggering conditions have been constructed for each event-triggered subsystem, where the triggering condition related to system stability is based on the states of the local subsystem and its neighbors, the other one built for recursive feasibility is only based on the state deviation between the current actual state and its predicted value of the local subsystem. The dual-mode strategy is adopted and an event-triggered DMPC algorithm is finally developed. Furthermore, the recursive feasibility and stability of the global system have been proved.

References

1. Scattolini, R. (2009). Architectures for distributed and hierarchical model predictive control-a review. *Journal of Process Control*, *19*(5), 723-731.
2. Vazquez, S., Leon, J. I., Franquelo, L. G., Rodriguez, J., Young, H. A., Marquez, A., & Zanchetta, P. (2014). Model predictive control: A review of its applications in power electronics. *IEEE Industrial Electronics Magazine*, *8*(1), 16-31.
3. Ferrara, A., Sacone, S., & Siri, S. (2016). Design of networked freeway traffic controllers based on event-triggered control concepts. *International Journal of Robust and Nonlinear Control*, *26*(6), 1162-1183.
4. Maestre, J. M., & Negenborn, R. R. (Eds.). (2014). *Distributed model predictive control made easy* (Vol. 69). Springer.
5. Song, Y., Wei, G., & Liu, S. (2016). Distributed output feedback MPC with randomly occurring actuator saturation and packet loss. *International Journal of Robust and Nonlinear Control*, *26*(14), 3036-3057.
6. Li, H., & Shi, Y. (2013). Robust distributed model predictive control of constrained continuous-time nonlinear systems: A robustness constraint approach. *IEEE Transactions on Automatic Control*, *59*(6), 1673-1678.
7. Dunbar, W. B. (2007). Distributed receding horizon control of dynamically coupled nonlinear sys-

tems. *IEEE Transactions on Automatic Control*, *52*(7), 1249-1263.

8. Li, S., Zheng. Y., & Lin, Z. (2014). Impacted-region optimization for distributed model predictive control systems with constraints. *IEEE Transactions on Automation Science and Engineering*, *12*(4), 1447-1460.
9. Wang, P., & Ding, B. (2013). Distributed receding horizon control for dynamically coupled large scale systems. *IFAC Proceedings Volumes*, *46*(13), 254-259.
10. Farina, M., & Scattolini, R. (2012). Distributed predictive control: A non-cooperative algorithm with neighbor-to-neighbor communication for linear systems. *Automatica*, *48*(6), 1088-1096.
11. Baldivieso Monasterios, P., Hernandez Vicente, B., & Trodden, P. A. (2017, July). Nested distributed model predictive control. In *IFAC-PapersOnLine* (Vol. 50, No. 1). Elsevier.
12. Trodden, P. A., & Maestre. J. M. (2017). Distributed predictive control with minimization of mutual disturbances. *Automatica*, *77*, 31-43.
13. Yin, X., Yue, D., & Hu, S. (2015). Model-based event-triggered predictive control for networked systems with communication delays compensation. *International Journal of Robust and Nonlinear Control*, *25*(18), 3572-3595.
14. Li, H., & Shi, Y. (2014). Event-triggered robust model predictive control of continuous-time nonlinear systems. *Automatica*, *50*(5), 1507-1513.
15. Hashimoto, K., Adachi, S., & Dimarogonas, D. V. (2015). Distributed aperiodic model predictive control for multi-agent systems. *IET Control Theory and Applications*, *9*(1), 10-20.
16. Zou, Y., Su, X., & Niu, Y. (2016). Event-triggered distributed predictive control for the cooperation of multi-agent systems. *IET Control Theory and Applications*, *11*(1), 10-16.
17. Eqtami, A., Heshmati-Alamdari, S., Dimarogonas, D. V., & Kyriakopoulos, K. J. (2013). A selftriggered model predictive control framework for the cooperation of distributed nonholonomic agents. In *2013 IEEE 52th Conference on Decision and Control (CDC)* (pp. 7384-7389). IEEE.
18. Kothare, M. V., Balakrishnan, V., & Morari, M. (1996). Robust constrained model predictive control using linear matrix inequalities. *Automatica*, *32*(10), 1361-1379.
19. Betti, G., Farina, M., & Scattolini, R. (2014). Realization issues, tuning, and testing of a distributed predictive control algorithm. *Journal of Process Control*, *24*(4), 424-434.
20. Branicky, M. S. (1998). Multiple Lyapunov functions and other analysis tools for switched and hybrid systems. *IEEE Transactions on Automatic Control*, *43*(4), 475-482.
21. Mi, X., Zou, Y., & Li, S. (2018). Event-triggered MPC design for distributed systems toward global performance. *International Journal of Robust and Nonlinear Control*, *28*(4), 1474-1495.